International Max Planck Research School (IMPRS)
for Maritime Affairs
at the University of Hamburg

More information about this series at
http://www.springer.com/series/6888

Hamburg Studies on Maritime Affairs
Volume 31

Edited by

Jürgen Basedow
Monika Breuch-Moritz
Peter Ehlers
Hartmut Graßl
Tatiana Ilyina
Florian Jeßberger
Lars Kaleschke
Hans-Joachim Koch
Robert Koch
Doris König
Rainer Lagoni
Gerhard Lammel
Ulrich Magnus
Peter Mankowski
Stefan Oeter
Marian Paschke
Thomas Pohlmann
Uwe Schneider
Detlef Stammer
Jürgen Sündermann
Rüdiger Wolfrum
Wilfried Zahel

Elke Ludewig

On the Effect of Offshore Wind Farms on the Atmosphere and Ocean Dynamics

 Springer

Elke Ludewig
Theoretical Oceanography
Institute of Oceanography
Hamburg
Germany

Dissertation zur Erlangung der Doktorwürde an der Fakultät für Mathematik, Informatik und Naturwissenschaften
Fachbereich Geowissenschaften der Universität Hamburg
Vorgelegt von: Elke Ludewig
Erstgutachter: PD Dr. Thomas Pohlmann
Zweitgutachter: Prof. Dr. Heinke Schlünzen
Tag der Disputation: 17. Januar 2014

ISSN 1614-2462   ISSN 1867-9587 (electronic)
ISBN 978-3-319-08640-8   ISBN 978-3-319-08641-5 (eBook)
DOI 10.1007/978-3-319-08641-5
Springer Cham Heidelberg New York Dordrecht London

Library of Congress Control Number: 2014955145

Printed on acid-free paper

Springer is part of Springer Science+Business Media (www.springer.com)

"Die Welle beugt sich jedem Winde gern."
Johann Wolfgang von Goethe, Faust II,
Vers7853/Thales

Preamble

The study of the effects of renewable energies on the earth system is quite new and asks for analysis. Especially the renewable energy wind plays a key role in Europe, and so I was very glad to work within that field of research during my dissertation. This book comprises the results of my dissertation, which was created and supported at the University of Hamburg and by the International Max Planck Research for Maritime Affairs (IMPRS-MA).

At this point, I want to use the chance to register some attendants supporting me and my work in the 32 months of my dissertation's progress.

Primarily, I have to thank my first adviser and tutor PD Dr. Thomas Pohlmann for his collaboration, for his offer of the scholarship at the *International Max Planck Research School for Maritime Affairs*, and especially for his professional support.

I have to thank my second adviser, Prof. Dr. Schlünzen, and the members of my examination committee, Prof. Dr. Gajewski, Prof. Dr. Backhaus, and Prof. Dr. Burchard, for their expenditure of time for my promotion proceeding.

Particularly, I have to mention M. Linde. Thank you for a well collaboration and expenditure of time by simulations, data, and questions regarding METRAS.

To the BSH, the WEGA cruise, and A. Schneehorst, a big thank you for an interesting cruise to *alpha ventus* and the measurement campaign that ends in a really wonderful revealing data set and a friendly collaboration.

A thank you is to be addressed to the TO work group, which provides a nice working atmosphere, and thank you to the IMPRS-MA for providing an insight into the law side of the maritime affairs and for supporting this book.

A thank goes also to the department of Informatics Scientific Computing of Hamburg University, especially H. Lenhart, for their interest in my work. I am really curious and looking forward to the simulations considering the OWF effect on the North Sea's ecosystem.

Last but not least, I want to mention my family and want to thank them for supporting my academic studies, my conference, and summer school trips, and you are always open to me.

Hamburg, Germany   Elke Ludewig
May 2014

Abstract

Nowadays, renewable energy resources play a key role in the energy supply discussion, and especially a heightened interest in wind energy induces intensified installation of wind farms. In the course of a larger demand of renewable energy, offshore wind farms (OWFs) gain increasingly in popularity since over-sea yields are larger and more reliable than over land. In this context, Germany adopts the position of a pioneering nation due to its national interurban offshore wind energy program comprising an intensified construction of wind turbines in the Baltic Sea and, mainly, North Sea. Against this background, it becomes particularly urgent to inquire whether and to what extent such OWF expansion affects our oceans and local climates.

OWFs excite wind speed reduction downstream of wind farms, the so-called wake effect, which impacts the atmosphere's boundary layer; locally disturbs the wind characteristics; and in turn affects ocean dynamics. To study the whole complex in more detail, investigations comprise *model simulations and measurements*. Used models are the atmosphere model *METRAS* (*ME*soscale *TRA*nsport and *S*tream model) and the ocean model *HAMSOM* (*HAM*burg *S*helf *O*cean *M*odel). METRAS simulations were generated in collaboration with and by courtesy of the Institute for Meteorological of the University of Hamburg. These METRAS data represent the meteorological forcing for simulations of the ocean. Measurements were taken around German test wind farm alpha ventus supported by the German Federal Maritime Service (BSH).

Analysis regarding OWF effect on the atmosphere and ocean comprises two main studies to determine possible OWF effects and their physical appearance in theory and to estimate possible future integrated changes of the North Sea's marine system based on the offshore construction plan for 2030. Investigations consider different amounts of wind turbines, wind speeds and directions, ocean depths, and forcing assumptions. Model results and measurements show a reasonable agreement supporting the principle validity of the used model approach.

Main results of this study show significant dynamical changes, including a *wind speed reduction* downstream of OWF up to 70 % over an area being 100 times larger than OWF itself, an evolving *dipole structure of the sea surface elevation* around OWFs, and *up- and downwelling cells* with a horizontal extension of approximately 30×30 km, spanning the whole ocean depth. The connected vertical velocities reach magnitudes of 3–4 m/day. In turn, these vertical motions introduce changes in stratification of temperature and salinity, which results in a maximal *excursion of the thermocline* by possibly 10 m. Hence, it can be concluded that offshore wind farms cause an intensified vertical mixing in the ocean, which may result in a fundamental change of the North Sea's ecosystem.

Zusammenfassung

Heutzutage spielen erneuerbare Energien eine Schlüsselrolle in der Diskussion zukünftiger Energieversorgung. Besonders ein verstärktes Interesse an Windenergie bewirkt einen intensiven Ausbau von Windfarmen. Im Zuge der erhöhten Nachfrage an erneuerbaren Energien gewinnen Offshore Windfarmen (OWFs). vermehrt an Popularität, zumal auf See größere und vorallem zuverlässig Erträge erzielt werden können. In diesem Zusammenhang nimmt Deutschland, infolge des nationalen Offshore Windenergieausbauprogramm, welches eine intensive Errichtung von Windkraftanlagen in der Ostsee und besonders in der Nordsee beinhaltet, eine Vorreiterrolle ein. Vor diesem Hintergrund ist es sehr bedeutsam abschätzen zu können, ob und in welchem Ausmaß ein solcher offshore Windfarmausbau unsere Meere und lokale Klima beeinflusst.

OWFs bewirken eine Reduktion der Windgeschwindigkeit in Windrichtung hinter der Windfarm. Diese Reduktion der Windgeschwindigkeit wird als Wake-Effekt bezeichnet. Der Wake-Effekt beeinflusst die atmosphärische Grenzschicht und lokal die Windeigenschaften, was wiederum Auswirkungen auf die Ozeandynamik zur Folge hat. Um den ganzen komplexen Sachverhalt der OWF Auswirkungen zu erfassen, wurden *Modellsimulationen und Messungen* für die Analyse herangezogen. Bei den verwendeten Modellen handelt es sich um das atmosphärische Modell *METRAS* (*ME*soskaliges *TRA*nsport und *S*trömungsmodell) und das Ozeanmodell *HAMSOM* (*Ham*burg Schelfmeer/*O*zean *M*odell). Simulationen mit METRAS wurden in Zusammenarbeit mit dem Meteorologischen Institut der Universität Hamburg erstellt und freundlicherweise dieser Arbeit zur Verfügung gestellt. Diese mit METRAS simulierten Daten dienen als meteorologischen Antrieb der Ozeansimulationen. Messungen wurden rund um den deutschen Testwindpark alpha ventus genommen. Die Messkampagne wurde vom Bundesamt für Schifffahrt und Hydrographie (BSH) unterstützt.

Analysen des OWF-Effekts auf Atmosphäre und Ozean umfassen zwei Hauptstudien, um den möglichen OWF-Einfluss und dessen physikalisches Auftreten theoretisch zu erfassen und mögliche Änderungen des marinen Systems der Nordsee bedingt durch den geplanten Offshore Ausbauplan für 2030. Untersuchungen berücksichtigen verschiedene Mengen und Anordnungen von

Windturbinen, Windgeschwindigkeiten und Modellantrieben. Modellergebnisse und Messungen zeigen eine angemessene Übereinstimmung, die den gewählten Modellansatz und prinzipielle Annahmen bestätigen.

Hauptergebnisse dieser Arbeit bezeugen signifikante dynamische Änderungen, zum einen in Bezug auf das Windfeld mit einer Reduzierung der Windgeschwindigkeit über ein Gebiet, welches hundertmal größer ist als die Windfarmfläche, bis 70 % und zum anderen in Bezug auf den Ozean durch das Auftreten von Wasserstandänderung mit Dipolstruktur, Up- und Downwellingzellen mit einer horizontalen Ausdehnung von rund 30 × 30 Kilometer über die ganze Meerestiefe. Die damit verknüpften vertikalen Geschwindigkeiten erreichen drei bis vier Meter pro Tag und bewirken eine Änderung in der Ozeanschichtung von Temperatur und Salzgehalt mit einer Auslenkung der Thermoklinen um 10 m rund um den OWF. Daher muss man davon ausgehen, dass OWFs intensives vertikales Mischen verursachen, welches eventuell Änderungen im Ökosystem der Nordsee bewirkt.

Contents

Abbreviations

Variables

ζ	Surface elevation [m]
gwind	Geostrophic wind
p	Pressure [Pa]
SST	Sea surface temperature [°C]
ug	Geostrophic wind [m/s]
uv10	Horizontal wind field in 10 m [m/s]
velc.u	Ocean's velocity component u [m/s]
velc.v	Ocean's velocity component v [m/s]
velc.w	Ocean's velocity component w [m/s]
velh	Ocean's horizontal velocity field

Abbreviations in Analysis

BTM	HAMSOM simulations in barotropic mode
F*	Forcing fields (F01, F02, F03, F04)
HD*	Depth of ocean (HD60, HD30) [m]
OWF	Offshore wind farm
OWFr	Simulations considering wind turbines
REFr	Reference without wind turbines
src*	Source code manipulation
T*	Wind turbine number (T012, T048, T080, T160, T8590) [#]
TOS	Type of simulation (TOS-01, TOS-2)
TS*	Temperature and salinity (TS01, TS02, TS03)
UG*	Prescribed geostrophic wind speed (UG5, UG8, UG16) [m/s]
wd*	Wind direction (N, NE, E, SE, S, SW, W, NW, N) [°]

Model and Data

ADCP   Acoustic Doppler Current Profiler
CTD   Conductivity–Temperature–Depth
ECMWF   European Centre for Medium-Range Weather Forecasts
HAMSOM   Hamburg Shelf Ocean Model
METRAS   Mesoscale transport and stream model
NOAA   National Oceanic and Atmospheric Administration
WOA   World Ocean Atlas

Institutions and Additionals

BMU   Bundesministerium für Umwelt, Naturschutz und Reaktorsicherheit
BMWi   Bundesministerium für Wirtschaft und Technologie
BSH   Federal Maritime and Hydrographic Agency
BWE   Bundesverband Wind Energie
dena   Deutsche Energie-Agentur GmbH
EEZ   Exclusive economic zone
IMPRS   International Max Planck Research School

Chapter 1
Introduction

Presently, we are living in an era of a turnaround in energy policy with strong interests in renewable energies, focusing on wind energy. Increased incident issues on living conditions based on climate change, on one hand, and the fear of nuclear power hazards, on the other hand, bearing problems of nuclear waste and common protests again nuclear energy, result in intense political discussions on applying renewable energies as main energy source. Especially Germany official heralds the energy turnaround in 2010. On September 28, 2010, the German Federal Cabinet enacted the so-called, in German, Energiekonzept (translation: energy concept). In this concept, the Federal Government postulates the aim to form Germany as one of the most energy-efficient and most environmentally friendly national economy in the near future by offering competitive energy prices and conserving the high-prosperity level of Germany. The main aims of this procedure are the phaseout of nuclear energy and the reduction of greenhouse gases by 40 % till 2020 and about 80 % till 2050 (BMWi and BMU 2012). At this juncture, renewable energies, notably wind energy, play an important role in reaching such aims. The percentage of renewable energy electricity generation on gross electricity consumption shall add up to 50 % in 2030 and 80 % in 2050 (BMWi and BMU 2012), whereas the German Federal Government highlights the importance of offshore wind energy as a major element for an environmentally friendly, reliable, and affordable energy supply (BMWi and BMU 2012). Additional offshore is favored due to geographical usable areas, a higher reliability due to consequent high wind speeds over ocean supported by less friction than for onshore structures, and even less political opposition of the population by avoiding the so-called Nimby-Effect, an effect describing shadow and noise disruption realizing health effects for humans. Taking for granted these facts, Germany commands a huge area in the North and Baltic Seas. Accordingly, the development goal of offshore energy is ambitious—a minimum of 25 GW of offshore energy supply till 2030 in the North and Baltic Seas, which accords 15 % of Germany's total energy demand. Based on year 2012, counting an energy demand of around 617.6 TWh, partitioned in 19.1 % stone coal,

© Springer International Publishing Switzerland 2015

E. Ludewig, *On the Effect of Offshore Wind Farms on the Atmosphere and Ocean Dynamics*, Hamburg Studies on Maritime Affairs 31,
DOI 10.1007/978-3-319-08641-5_1

25.7 % brown coal, 11.3 % natural gas, 5.7 % mineral oil and others, 22 % renewable energy (wind, biomass, water, photovoltaic, biogenic garbage), and 16.1 % nuclear energy (BMWi and BMU 2012), offshore wind energy can be a replacement for nuclear energy.

But Germany is not alone in using wind energy. There exist attractive locations for the wind industry, used in the near future, worldwide. The Global Wind Energy Council and Greenpeace International present in their fourth edition of the *Global Wind Energy (GWEC) Outlook 2012* (GWEC and Greenpeace International 2012) in three different scenarios the total installed capacity of worldwide installed wind farms by 917,798 MW up to 2,541,135 MW for the year 2030. Based on GWEC's statistics of mid-2013, there has been 4,630 MW of offshore wind power installed globally, representing about 2 % of the total installed wind power capacity. More than 90 % of it is installed in northern Europe alone, and most of the rest is in two demonstration projects off China's east coast. However, there are also great expectations placed for major deployment elsewhere; governments and companies in Japan, Korea, the United States, Canada, Taiwan, and even India have shown enthusiasm for developing offshore in their waters. According to the more ambitious projections, a total of 80 GW of offshore wind could be installed by 2020 worldwide, with three-quarters of this in Europe (GWEC and Greenpeace International 2012).

Political energy plans show that Europe prefers offshore wind farming, like other countries having access to the ocean. That underlines that, in the future, wind power will increase worldwide, which leads to scientific questions dealing with the effects of wind turbines on our environment and atmospheric and oceanic surroundings. So what significance does wind farming have for us? What will happen if we establish wind farms near our coasts? To clarify the impact, one has to take into account that the term 'wind farm' can be defined as a power plant using a congeries of wind turbines to generate a high total power of electricity. In the case of Germany, this again means the construction of diverse offshore wind farms (OWFs) in Germany's exclusive economic zone (EEZ), a huge area that can be filled with hundreds of wind turbines. Here, such a development will change the North Sea's appearance and leads to the question on what impact such a shift can have on the atmosphere and ocean considering the energy transformation of atmospheric energy over mechanical to electrical energy.

The effect of wind turbines can be treated in different ways. Done scientific studies are separated into industrial and technical aspects, analyzed effects on the atmosphere, and analyses of biosphere, ecosystem, and medical impacts.

The *technical sector* concentrates on the potential of energy, the arrangement of wind turbines in a field, the size and form of rotor blades, the power of turbines, and duration of life, treated in Jenkins (1993), Mosetti et al. (1994), Sutherland and Mandell (1996), Polinder et al. (2005), Castro Mora et al. (2007), and Lackner and Elkinton (2009), just to list a handful of examples across the last decades. Based on industrial impacts and profit thinking, these topics are well analyzed and optimized but are still in active research for more optimizing and to aim reduction of costs.

Besides technical analysis, some *studies deal with the effects on biosphere and ecosystem and human life,* for example Zettler and Pollehne (2006), Lange et al. (2010), Nunneri et al. (2008), and Wolsink (2000).

These studies underline issues regarding beards, bats, sea mammals or lobsters, and other sea animals, as well as noise and shadow effects (Nimby-Effect) bothering humans.

If we refrain from the medical and biosphere causes of wind turbines and have a closer look at other studies regarding wind turbines, then these studies concentrates on the effect of wind turbines on their surroundings. The focus of these *studies* is the *change in wind field and energy and their effects on the atmosphere,* like changes in temperature and wind field on *higher scales* (Baidya Roy and Traiteur 2010; Christiansen and Hasager 2005; Hasager et al. 2013; Zhou et al. 2012) and *small scales* (Jimenez et al. 2007; Porté-Agel et al. 2011; Wu and Porté-Agel 2010; Lu and Porté-Agel 2011).

There exist studies on how strong a wind farm can influence meteorological situations and *changing weather* (Fiedler and Bukovsky 2011; Fitch et al. 2012; Baidya Roy 2004; Kirk-Davidoff and Keith 2008; Keith et al. 2004).

Experiments were done dealing with the question as to what happens to *global energy distribution* by demanding big wind farms everywhere (Wang and Prinn 2010). The overall aspects of these studies are a reduction in wind speed behind wind farms in wind direction, the so-called wind wake, a mostly cooling at offshore and warming at onshore, as well as possible dynamical changes in the atmosphere due to the wind wake.

Studies dealing with dynamical effects on the ocean, like this dissertation, are quite new and rarely documented. First, Broström (2008) indicates a change on sea surface elevation due to wind farms, which is even documented in Paskyabi and Fer (2012).

Nerge and Lehnhart picked up Broström's concept. Their results are summarized in the LOICZ 2010 report, which shows the effect of wind farms on oceans in various scientific areas. Here, it becomes clear that offshore wind farms have an important influence on oceans. That is why this book uptakes Nerge and Lehnhart first results to engross and discuss important physical aspects of offshore wind farm effects roundly.

This book concentrates on whether and in what manner the dynamics will change if we extract energy from the atmosphere over a big areal domain. Based on listed known studies and considering the political situation, the core question of this book is analyzed in adaption to Germany's situation in the North Sea. But all results can be also associated with other coastal regions. The focusing on Germany and the North Sea follows practicable reasons, the strong interests of Germans in this subject, as well as the work's frame, including financial and scientific support by German institutions, namely the IMPRS for Maritime Affairs, the University of Hamburg, and the Federal Maritime and Hydrographic Agency (BSH).

In this connection, the aims are to analyze and explain the dynamical effects that offshore wind farms have on oceans due to the mentioned wind reduction and wake production and to provide this information for further studies with an economical background to answer the question of the possibility of reef building, mussel farming, and other ecological changes in the future, as well as to support additional projects in that field. To get closer to those oceanic questions, it is necessary to include the atmosphere, that is, why here also common influences of wind farms on the atmosphere are presented even though they are now easily found in literature.

The analysis of offshore wind farm effects on the ocean and atmosphere comprises model simulations, as well as measurements, and is organized as follows.

To become acquainted with the book's topic, Chap. 2 gives an introduction into wind energy and, especially, offshore wind energy. Chapter 3 explains used data and models and gives an overview of applied methods. The heart of this work is Chaps. 4–6, comprising three analyses. Chapter 4 describes the effect of wind farms on the atmosphere based on theoretical assumptions, which spans explanation of forcing for ocean modeling. Chapter 5 presents the effect of OWF on an idealized ocean box, including different wind forcing, analysis of physical ocean processes triggered by OWFs, and model evaluation with measurements. The analysis in Chap. 6 gives an insight into the future of the German Bight regarding the demand of offshore wind farms in the North Sea. Finally, Chap. 7 summarizes and gives an outlook.

References

Baidya Roy S (2004) Can large wind farms affect local meteorology? J Geophys Res 109:D19101. doi:10.1029/2004JD004763

Baidya Roy S, Traiteur JJ (2010) Impacts of wind farms on surface air temperatures. Proc Natl Acad Sci U S A 107:17899–17904. doi:10.1073/pnas.1000493107

BMWi, BMU (2012) Erster Monitoring-Bericht "Energie der Zukunft." BMWi und BMU Report, Öffentlichkeitsarbeit, pp 1–132

Broström G (2008) On the influence of large wind farms on the upper ocean circulation. J Marine Syst 74:585–591

Castro Mora J, Calero Barón JM, Riquelme Santos JM, Burgos Payán M (2007) An evolutive algorithm for wind farm optimal design. Neurocomputing 70:2651–2658. doi:10.1016/j.neucom.2006.05.017

Christiansen MB, Hasager CB (2005) Wake studies around a large offshore wind farm using satellite and airborne SAR. In: 31st international symposium on remote sensing of environment, St. Petersburg (Russian Federation)

Fiedler BH, Bukovsky MS (2011) The effect of a giant wind farm on precipitation in a regional climate model. Environ Res Lett 6:045101. doi:10.1088/1748-9326/6/4/045101

Fitch AC, Olson JB, Lundquist JK et al (2012) Local and mesoscale impacts of wind farms as parameterized in a mesoscale NWP Model. Mon Weather Rev 140:3017–3038. doi:10.1175/MWR-D-11-00352.1

GWEC, International G (2012) Global wind energy outlook, 2012. REPORT, pp 1–52

Hasager C, Rasmussen L, Peña A et al (2013) Wind farm wake: the horns rev photo case. Energies 6:696–716. doi:10.3390/en6020696

Jenkins N (1993) Engineering wind farms. Power Eng J 7:53–60

Jimenez A, Crespo A, Migoya E, Garcia J (2007) Advances in large-eddy simulation of a wind turbine wake. J Phys Conf Ser 75:012041. doi:10.1088/1742-6596/75/1/012041

Keith DW, DeCarolis JF, Denkenberger DC et al (2004) The influence of large-scale wind power on global climate. Proc Natl Acad Sci USA 101:16115–16120

Kirk-Davidoff DB, Keith DW (2008) On the climate impact of surface roughness anomalies. J Atmos Sci 65:2215–2234. doi:10.1175/2007JAS2509.1

Lackner MA, Elkinton CN (2009) An analytical framework for offshore wind farm layout optimization. Wind Eng 31:17–31. doi:10.1260/030952407780811401

Lange M, Burkhard B, Garthe S, Gee K (2010) Analyzing coastal and marine changes: offshore wind farming as a case study. LOICZ Res Stud 36:212

Lu H, Porté-Agel F (2011) Large-eddy simulation of a very large wind farm in a stable atmospheric boundary layer. Phys Fluids 23:065101. doi:10.1063/1.3589857

Mosetti G, Poloni C, Diviacco B (1994) Optimization of wind turbine positioning in large windfarms by means of a genetic algorithm. J Wind Eng Ind Aerodynamics 51:105–116. doi:10.1016/0167-6105(94)90080-9

Nunneri C, Lenhart HJ, Burkhard B, Windhorst W (2008) Ecological risk as a tool for evaluating the effects of offshore wind farm construction in the North Sea. Reg Environ Chang 8:31–43. doi:10.1007/s10113-008-0045-9

Paskyabi MB, Fer I (2012) Upper ocean response to large wind farm effect in the presence of surface gravity waves. Energy Procedia 24:245–254. doi:10.1016/j.egypro.2012.06.106

Polinder H, de Haan S, Dubois MR (2005) Basic operation principles and electrical conversion systems of wind turbines. EPE J 15:43

Porté-Agel F, Wu Y-T, Lu H, Conzemius RJ (2011) Large-eddy simulation of atmospheric boundary layer flow through wind turbines and wind farms. J Wind Eng Ind Aerodynamics 99:154–168. doi:10.1016/j.jweia.2011.01.011

Sutherland HJ, Mandell JF (1996) Application of the US high cycle fatigue data base to wind turbine blade lifetime predictions. In: Proceeding of Energy Week, AMSE

Wang C, Prinn RG (2010) Potential climatic impacts and reliability of very large-scale wind farms. Atmos Chem Phys 10:2053–2061. doi:10.5194/acp-10-2053-2010

Wolsink M (2000) Wind power and the NIMBY-myth: institutional capacity and the limited significance of public support. Renew Energy 21:49–64. doi:10.1016/S0960-1481(99)00130-5

Wu Y-T, Porté-Agel F (2010) Large-eddy simulation of wind-turbine wakes: evaluation of turbine parametrisations. Boundary-Layer Meteorol 138:345–366. doi:10.1007/s10546-010-9569-x

Zettler ML, Pollehne F (2006) The impact of wind engine constructions on benthic growth patterns in the western Baltic. In: Köller J, Köppel J, Peters W (eds) Offshore wind energy research on environmental impacts. Springer, Berlin, pp 201–222

Zhou L, Tian Y, Roy SB et al (2012) Impacts of wind farms on land surface temperature. Nat Clim Chang 2:1–5. doi:10.1038/nclimate1505

Chapter 2
Renewable Energy Wind

This book deals with the current most important and effective renewable energy, the wind energy. Countries all over the world want to place wind turbines into nature for producing 'green energy.' In the last decades, renewable energies were fixed in our society, aiming on protecting resources and providing a new greener and, hence, more healthful life. Common press coverage delivers information on the status and development in this field. Therefore, this chapter gives an insight into the renewable wind energy and its current state and technique. Assuming the crucial point that this book concentrates on offshore wind farms, this chapter finishes with a description of Germany's wind farming in the North Sea, including a description of the first German test wind farm, called *alpha ventus*, and a possible future offshore expansion.

2.1 Utilization of Wind Energy: Historical and Technical Background

Transforming wind power into mainly mechanical energy has been established in our civilization since centuries ago. A nice overview of the history of wind turbine is given by Hau (2008).

The first historical source of a windmill's existence dates back to 644 AD and describes a windmill with vertical rotation axes placed in the Persian–Afghan border district and used for grinding grain. Across the centuries, references arise in China of windmills that were used to drain paddy fields. Windmills with a horizontal axis of rotation were developed in Europe, may independently of the construction of vertical-axis windmills' in the Normandy. Over the centuries, windmills have been technically improved and have spread over Europe and Russia. So various types of windmills arise—the ancestors of today's wind turbines. Windmills that are similar to the turbines of today were developed in the nineteenth

© Springer International Publishing Switzerland 2015
E. Ludewig, *On the Effect of Offshore Wind Farms on the Atmosphere and Ocean Dynamics*, Hamburg Studies on Maritime Affairs 31,
DOI 10.1007/978-3-319-08641-5_2

century in the mid-western United States. They were mainly used for water pumping. Proposed by the Danish government, which is looking for possibilities to supply rural areas with electricity, Danish professor Poul La Cour built in 1891 an experimental wind turbine driving a dynamo in Askov, Denmark. While in the past windmills were mainly used for transforming wind energy into mechanical energy, generating electricity is prioritized today.

Enhancing Poul La Cour's construction, nowadays wind turbines comprise a huge development in engineering. A sophisticated technique transforms wind energy into electricity by controlling the blades and the turbine's conditions, focusing on high efficiency. Over the years, wind turbines of a high technical standard were developed to transform wind energy into electrical power.

2.1.1   Today's Wind Turbines

The main concept of windmill nowadays is a tower, mostly three rotor blades and a turbine in a nacelle (Morris 2006). Such a wind turbine is illustrated in Fig. 2.1.

Today, various turbines are available for different conditions and power supplies. The first turbines dealt with energy of 2 megawatts (MW). Now 5-MW turbines are common, but the turbines will be developed by ten, 15 (promoted by Spanish companies Gamesa, Iberdrola, and Acciona in 2010), and maybe more in future times. The diameter of the rotor started with a couple of meters; 80 to a 100 m are currently used, and bigger rotor blades up to 200 m are in the testing phase (Vestas Wind Systems 2013).

With regard to energy production, there exist two different styles of nacelles—construction with gear mechanism and one without gear mechanism. As regards construction with gear mechanism, the power is lead from a rotating motion of the rotor through a driveshaft and gear in the dynamo. A dynamo can only work with a high driving speed, which cannot be supported by wind turbines. This issue is solved with the help of the gear, which transforms power with low driving speed and high turning moment into power with high driving speed and low turning moment. In contrast, constructions without gear have an advantage of less machine components. This means less rotating materials, hence less maintenance works. Therefore, they are proposed for offshore turbines where maintenance is complicated, time and cost intensive, and weather dependent.

Such constructions offer a synchronous generator activated by a permanent magnet. They transform the rotor motion directly into electricity. In sum, nearly 45 % of wind energy can be transformed into electrical power (BWE 2014). The rest is lost during the transformation process. Around 41 % of wind energy cannot be extracted; the rest of 59 % is reduced by aerodynamic rotor losses, mechanical losses, and electrical losses through the driveshaft and generator (BWE 2014).

Wind turbines need a strong basement structure to fix them. On one hand, a rotating turbine leads to imbalance and vibration during rotation; on the other hand, wind and waves exert force on construction. While onshore turbines mostly fixed by

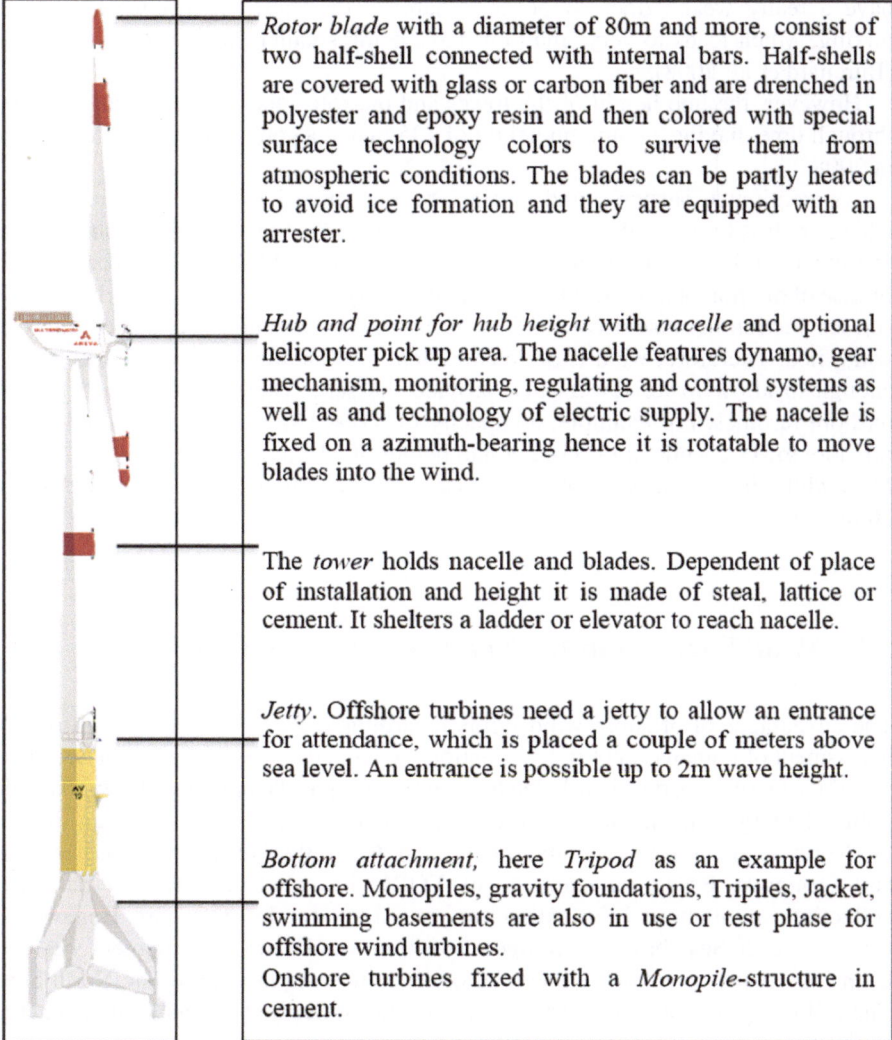

Rotor blade with a diameter of 80m and more, consist of two half-shell connected with internal bars. Half-shells are covered with glass or carbon fiber and are drenched in polyester and epoxy resin and then colored with special surface technology colors to survive them from atmospheric conditions. The blades can be partly heated to avoid ice formation and they are equipped with an arrester.

Hub and point for hub height with *nacelle* and optional helicopter pick up area. The nacelle features dynamo, gear mechanism, monitoring, regulating and control systems as well as and technology of electric supply. The nacelle is fixed on a azimuth-bearing hence it is rotatable to move blades into the wind.

The *tower* holds nacelle and blades. Dependent of place of installation and height it is made of steal, lattice or cement. It shelters a ladder or elevator to reach nacelle.

Jetty. Offshore turbines need a jetty to allow an entrance for attendance, which is placed a couple of meters above sea level. An entrance is possible up to 2m wave height.

Bottom attachment, here *Tripod* as an example for offshore. Monopiles, gravity foundations, Tripiles, Jacket, swimming basements are also in use or test phase for offshore wind turbines.
Onshore turbines fixed with a *Monopile*-structure in cement.

Fig. 2.1 Schematic illustration of a typical wind turbine that is used nowadays in offshore (onshore) wind farms. Illustration is based on a schema by Trianel Windkraftwerk Borkum GmbH & Co. KG

a monopole and a cement basement, offshore wind turbines can be fixed by various basement structures like tripods, jackets, monopiles, or swimming basements. Swimming basements are still in the testing phase (FLOATGEN project), while currently tripods are mostly used. In the case of offshore wind park *alpha ventus* (described in Sect. 5.4), tripods and jackets are tested. Piles will be ignored in this study due to the fact that the model resolution cannot resolve such small turbulences. But that assumption will not influence analysis because turbines themselves have a really weak effect on ocean dynamics; they just support weak vertical mixing, which leads to very small changes, being smaller than natural variability (Burchard et al. 2008).

However, the hub height of the tower and the rotor diameter show an increase through time, having a common height of 135 m and a common diameter of 126 m in 2008.

As mentioned, onshore wind turbines become bigger to catch stronger winds, which are less influenced and weakened by surface friction (through bushes, tress, and other surface constitutions), which again supports a higher energy production. In case of offshore wind farming, it is not necessary to counteract wind reduction by surface friction because ocean surface friction is less and no structures disturb the wind filed. Therefore, wind is quite consistent in offshore areas, as well as strong enough to allow lower towers that provide the same capacity like much higher onshore turbines. For example, a capacity of 3,000 kW can be reached by 80 m offshore towers, while at onshore, a hub height of 110 m is required (BWE 2014). Meanwhile, in dealing with offshore cases, hub heights of used towers are set to 80 m.

2.2 Wind Farming in the North Sea: Example Germany

The advantages of offshore wind farms regarding space, productivity, and minor harassment of humans support constructions of wind turbines in the North Sea. The use of the German part of North Sea for wind farming is the only possibility to reach political energy aims mentioned in the introduction of the book. Wind farming in the North Sea is used by the United Kingdom, the Netherlands, Belgium, Norway, and Denmark. Based on statistics of LORC Knowledge, a Danish database collecting all offshore information supervised by Lindoe Offshore Renewable Center, North Sea, has the strongest offshore wind farm development besides China. Germany cabs tap the potential of offshore wind farming in the North Sea. The following section will shortly define the North Sea and its availability for use of OWFs.

2.2.1 The North Sea

The North Sea is located in Europe, having coasts to Norway, Great Britain, France, the Netherlands, Germany, and Denmark. It is defined as a shallow shelf sea with a mean depth of 80 m and a maximum water depth that is in the Norwegian Trench of about 800 m (Sündermann and Pohlmann 2011; Mathis 2013). The North Sea has got a strong tide and is famous for its Wadden Sea, declared as UNESCO world culture heritage.

A summary of synoptically, hydrodynamic, and hydrographic conditions of the overall North Sea area is given by Mathis (2013) based on Otto et al. (1990), Rodhe (1998), OSPAR (2000), and Steele et al. (2009).

Most relevant information of the North Sea for this study is its stratification of the area within the so-called exclusive economic zone, which is defined under Sect. 2.2.2. The status reports of the German Federal Maritime and Hydrographic Agency (BSH) provide information about the System North Sea. The results of that reports show a wind statistic of mostly wind directions between southwesterly and westerly winds for the southern part of the North Sea. On average over all seasons, the geostrophic wind speed has been counted around 8.0 m/s since 2005. That value is calculated at position 5°E and 55°N, which is representative of the area between 0°–10°E and 50°–60°N (Loewe 2009).

Sea surface temperatures count 10 °C on average. Geographically, SSTs increase from northwest to southeast. Most areas of the German EEZ are thermally stratified in the summer with temperatures of 6–7 °C at bottom and 15–18 °C, sometimes 20 °C, at surface (Loewe 2009).

Large parts of the North Sea are water of Atlantic origin, having salinity concentrations greater than 35. Hence, highly salty waters of Atlantic origin in the West (35/34 psu) and low salty waters (lower than 32 psu) in the East being influenced by Baltic sea waters and continental fresh water input, like those from the rivers Rhine and Elbe, characterize the distribution of salinity concentration of the North Sea.

That information is used to set up an idealized ocean having North Sea conditions to analyze dynamical changes due to operating offshore wind farms.

2.2.2   Germany's Exclusive Economic Zone (EEZ)

States having access to oceans are able to build offshore wind farms. As a result of UNCLOS (United Nations Convention on the Law of the Sea), states are allowed to build such wind farms only in a special zone near their coasts. This zone is called exclusive economic zone. 'The exclusive economic zone is an area beyond and adjacent to the territorial sea, subject to the specific legal regime established in this part, under which the rights and jurisdiction of the coastal state and the rights and freedoms of other states are governed by the relevant provisions of this convention,' UNCLOS, Part V, Article 55.

Based on this definition, Germany's area for offshore wind farms in the North Sea is restricted. The area of Germany's EEZ and its ocean depth are depicted in Fig. 2.2. Germany's EEZ includes important shipping routes that are indefeasible. Besides these routes, a huge area within the EEZ is useable for offshore wind farming even in depths of 60 m.

Plans of offshore wind farms demanded within the EEZ are listed in the LOIZ research and studies number 36 (Lange et al. 2010).

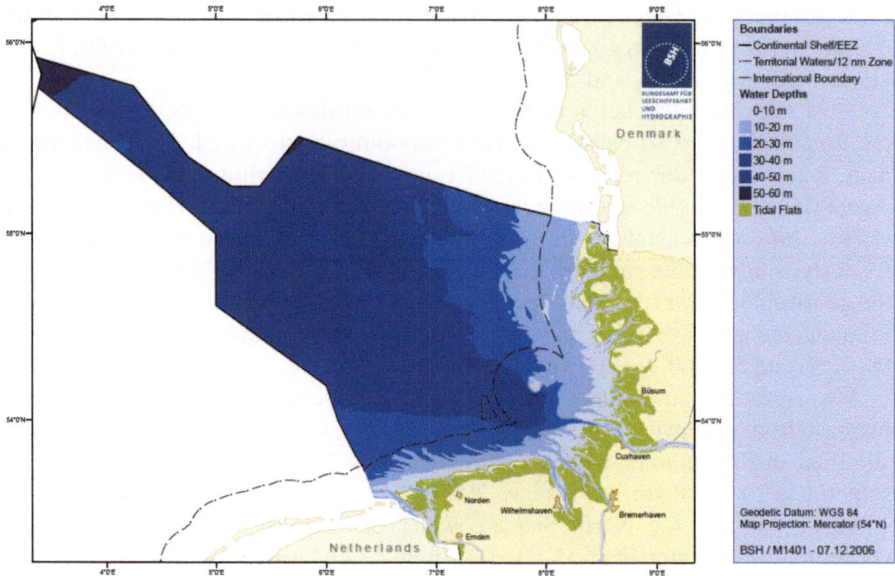

Fig. 2.2 Germany's part of the North Sea, continental shelf, exclusive economic zone. Colors show water depths with a maximum of 60 m. Area of EEZ in the North Sea spans around 28,600 km^2. Illustration obtained from BSH

These scenarios deal with different marine management perspectives of the EEZ. One of these scenarios, precisely an extreme scenario, is disposed and presented in this book in Chap. 6 to estimate possible changes of the North Sea in case of intense offshore wind farming.

Nowadays, four offshore wind farms are in operating state (BMU 2013): *alpha ventus* with 12 turbines, BARD Offshore 1 with 80 turbines, ENOVA Offshore Ems Emden and Hooksiel Offshore each with one turbine. Several more are approved, under construction, and under approval procedure. Figure 2.3 marks spaces within the EEZ that are planned and used for OWFs. BARD Offshore 1 lies 90 km west of Borkum; ENOVA nearshore, 0.1 km; and industry port Emden and Hooksiel nearshore, 0.5 km north of Wilhelmshaven.

Offshore wind farm alpha ventus has a special role in this book due to its status as Germany's first offshore wind farm and due to its attribute as 'test wind farm' and the focus of research. Details of this pregnant wind farm are given in the next section.

2.2.3 Wind Farm Alpha Ventus

The following information on alpha ventus is based on 'alpha ventus Fact Sheets,' whose publishing is supported by BMU (Bundesministerium für Umwelt,

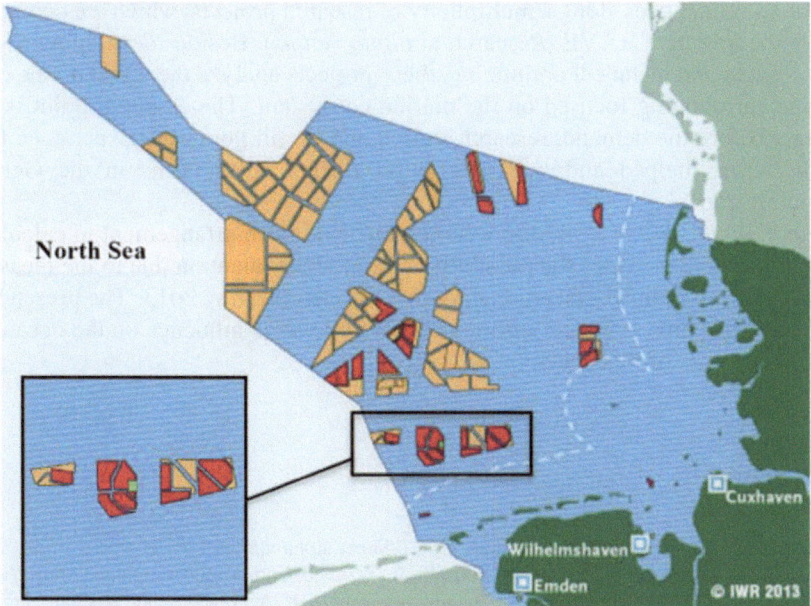

Fig. 2.3 Offshore wind farm projects in the North Sea, status 2013. *Green* selection identifies OWF in operating state, *red* selection marks OWFs that are approved/under construction, and *orange* areas are under approval procedures. *Zoomed area* shows a section north of island Borkum. *Green square* presents wind farm alpha ventus. The figure has a copyright by IWR and was taken from BMU (2013). It was changed only by adding and highlighting the zoomed area

Naturschutz und Reaktorsicherheit) (Engl.: Federal Ministry for the Environment, Nature Conservation and Nuclear Safety).

Alpha ventus is known as Germany's test wind park, which was developed by the consortium 'Deutsche Offshore-Testfeld und Infrastruktur GmbH & Co. KG (DOTI)' (translation: Germany's offshore test field and infrastructure GmbH & Co. KG) on June 2006 (Bartsch 2013). *Alpha ventus* was commissioned on April 2010. Hence, it was the first offshore wind park in Germany, with an investment sum of 250 million euros and 30 million euros supported by the BMU. The wind park is located 45 km north of the island Borkum, where the North Sea has a depth of around 30 m; see Fig. 2.2. The test field consists of 12 wind turbines; one research platform, named Fino1; and one relay station. Together, all of these wind turbines have got a capacity of 60 MW. Each wind turbine has got a rotor diameter of 116 m with a hub height of 90 m, a capacity of 5 MW, and a rated speed of 12.5 m/s. Speed limitations start at 3.5 m/s and ends at 25 m/s. The turbines are fixed on tripods and jackets (Fig. 2.3) and arranged in a 3 × 4 matrix with three rows in longitude and 4 rows in latitude. The location is mapped and treated in Sect. 5.4.

Alpha ventus goes along a multiplicity of research projects, which are comprised under the initiative RAVE (*Research at alpha ventus*). Besides developing a wind park system and technical optimizing, these projects analyze the effect on the close marine surrounding focused on the marine ecosystem. The essential point is that 2 years before the demand, research started and is still going on. Experience from *alpha ventus* helped and will help build further wind farms in the German economic zone.

Here, *alpha ventus* is used as a sample of wind farm arrangement in calculated simulations and provides the possibility of a model evaluation due to the measurements that were already taken around *alpha ventus* on May 2013. The presence of *alpha ventus* helps to collect information about OWF's influence on the ocean and confirms results of this study.

References

Bartsch C (2013) Fact-sheet alpha ventus. FACT-Sheet alpha ventus 1–8

BMU (2013) Offshore-windparks in Betrieb. In: www.offshore-windenergie.net. http://www.offshore-windenergie.net/windparks/windparks-in-betrieb. Accessed 2 Dec 2013

Burchard H, Huttmann F, Janssen F et al (2008) Effects of wind farm foundations on the water exchange between North Sea and Baltic Sea – a first careful assessment derived from the QuantAS-Off project. Projektbericht QuantAS-Off 27

BWE (2014) Windenergie – Technik. In: http://www.wind-energie.de/infocenter/technik/. Accessed 14 May 2014

Hau E (2008) Windkraftanlagen, 4th edn. Springer, Heidelberg

Lange M, Burkhard B, Garthe S, Gee K (2010) Analyzing coastal and marine changes: offshore wind farming as a case study. LOICZ Res Stud 36:212

Loewe P (2009) Bundesamt für Seeschifffahrt und Hydrographie – Berichte des BSH – System Nordsee. REPORTs by German Federal Maritime and Hydrographic Agency 1–270

Mathis M (2013) Projected forecast of hydrodynamic conditions in the North Sea for the 21st century. Dissertation Universität Hamburg 182

Morris N (2006) Wind power. Black Rabbit Books. ISBN:9781583409107

OSPAR (2000) Quality Status Report 2000: Region II: Greater North Sea. Marine, OSPAR Commission for the Protection of the Marine, London 136:25

Otto L, Zimmermann JTF, Furnes GK et al (1990) Review of the physical oceanography of the North Sea. Netherlands J Sea Res 26:161–238

Rodhe J (1998) The Baltic and North Seas: a process-oriented review of the physical oceanography. Coastal Segment (20° S). In: Robinson AR, Brink KH (eds) The sea, vol 11. Wiley, New York, pp 699–732

Steele JH, Thorpe SA, Turekian KK (2009) North Sea circulation. Encyclopedia of Ocean Sciences, 2nd edn, vol 4. Academic, London, pp 73–81

Sündermann J, Pohlmann T (2011) A brief analysis of North Sea physics. Oceanologia 53:663–689. doi:10.5697/oc.53-3.663

Vestas Wind Systems VG (2013) Vestas product brochures. In: www.vestas.com. http://www.vestas.com/en/media/brochures.aspx. Accessed 9 Dec 2013

Chapter 3
Models, Data, and Methodology

This work comprises two models and a couple of data, which are described here. The models used are mesoscale models named HAMSOM and METRAS, which were used to simulate the ocean and the atmosphere. Data comprise forcing data for the simulations, as well as done measurements. Additionally, this chapter explains the model setups and the methodology used to analyze the impact of offshore wind turbines on the atmosphere and the ocean.

3.1 Models

3.1.1 Hamburg Shelf Ocean Model

The **HAM**burg **S**helf **O**cean **M**odel (HAMSOM) is a numerical three-dimensional baroclinic hydrostatic dynamical model developed by Backhaus (1985). Here, a modified version, including improvements by Pohlmann (2006), is used. HAMSOM is well developed for sea shelf processes and tested in various studies of the North Sea and other shelf seas such as Backhaus and Hainbucher (1987), Carbajal (1993), Becker et al. (1999), Hainbucher and Backhaus (1999), and Huang et al. (1999), and several more.

HAMSOM is known as an accurate model simulating physical processes well and therefore features the best requirements for this study. Model basics are the primitive equation, a free surface. The horizontal and vertical grid spacing is defined in z-coordinates on the Arakawa C-grid (Arakawa and Lamb 1977). The model uses a semi-implicit scheme, instead of separating the internal and external modes, which are largely exempt from the stability criteria usually required for explicit formulations. A further detailed description of additional HAMSOM features is allocated in this study if necessary, for example in Sect. 5.2, which deals with a sensitivity study to analyze physical processes in the ocean.

© Springer International Publishing Switzerland 2015
E. Ludewig, *On the Effect of Offshore Wind Farms on the Atmosphere and Ocean Dynamics*, Hamburg Studies on Maritime Affairs 31,
DOI 10.1007/978-3-319-08641-5_3

HAMSOM's forcing includes data of wind, pressure, temperature, humidity, precipitation, and cloudiness with their origin of, on one hand, ECMWF (European Centre for Medium-Range Weather Forecasts) data and, on the other hand, forcing data also modeled by the atmosphere model METRAS.

3.1.2 Mesoscale Transport and Stream Model

The **ME**soscale **TRA**nsport and **S**tream (METRAS) model was developed by Schlünzen in 1988 and was complemented with a wind turbine parameterization by Linde et al. (2014). Hence, METRAS has implemented wind turbines and so is able to resolve changes due to wind turbines in the atmosphere. This study uses METRAS data as a meteorological forcing, which was modeled in collaboration with the Meteorological Institute (MI) of the University of Hamburg based on the need for ocean analysis. The METRAS data are a courtesy of the MI of the University of Hamburg, and the METRAS simulations were done by Marita Linde, whose Ph.D. includes atmospheric changes in the north of Germany due to OWFs.

Here, only wind turbine parameterization of METRAS is considered.

Wind Turbine Parameterization in METRAS
Facts about wind parameterization in METRAS are based on Linde et al. (2014) and personal correspondence with M. Linde (Ph.D. candidate at MI since 2011).

METRAS uses the actuator disc concept (ADC) for its wind turbine parameterization (Linde et al. 2014). This concept is based on Betz (1926a) constitute the rotor as an infinitesimal thin disc with a fixed rotor diameter and a midpoint at hub height and position of wind turbine. A schematic illustration of the concept of the ADC is shown in Fig. 3.1 after Betz (1926b), Mikkelsen (2003), and Linde et al. (2014).

It is assumed that an air package contains kinetic energy depending on its velocity. Far in front of a wind turbine, the air package is not influenced by the wind turbine and has the velocity v_1. Because of the extraction of kinetic energy, the flow velocity v_2 is reduced behind a wind turbine. This increases the pressure right in front of the rotor disc A' (Fig. 3.1). The parallel streamlines of laminar flow are spreading up. An air package that passes a small area A_1 far in front of the wind turbine passes a larger area A_2 far behind the wind turbine. The maximal thrust T_{max} is reached for $v_2 = 0$. Using these assumptions, a dimensionless thrust coefficient c_T can be formulated as the percentage of rotor thrust T' to maximum thrust with the air density ρ:

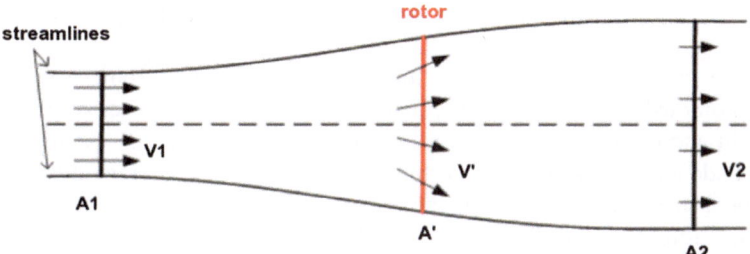

Fig. 3.1 Schema of actuator disc concept after Betz (1926b), courtesy of Linde et al. (2014). *Red line* marks rotor disc. For further details, please see text

$$c_T = \frac{T'}{T_{max}} = \frac{\frac{1}{2}\rho A'\left(v_1^2 - v_2^2\right)}{\frac{1}{2}\rho A' v_1^2} = 1 - \frac{v_2^2}{v_1^2} \tag{3.1}$$

The thrust coefficient c_T is a parameter given for each wind turbine type by the manufacturer's specifications, or it can be determined from field measurements.

The coefficient varies with the speed of wind. According to the definition of the thrust coefficient, the rotor thrust equation (Eq. 3.1) only depends on the wind speed of the undisturbed flow, the thrust coefficient, and the rotor area. The rotor area depends on the rotor diameter and is easily calculated. But the undisturbed wind speed has to be determined during model simulation.

$$T' = \frac{1}{2}\rho A'\left(v_1^2 - v_2^2\right) = c_T T_{max} = \frac{1}{2}c_T \rho A' v_1^2 \tag{3.2}$$

In METRAS, like in all mesoscale models, the horizontal grid size is typically large, compared to the size of the wind turbine rotor. It means that for a single wind turbine, the grid cell for the rotor and the reference position are the same. Also, several wind turbines might be located within one single grid cell. Therefore, a whole wind farm, like *alpha ventus* with 12 turbines, is only located within a few adjacent grid cells. Therefore, the wind reduction of several wind turbines superposes one large wind wake in the model. To adapt that issue, the wind counts ratable into affected grid box. That means a preprocessor mask consisted of an auxiliary grid allows the calculation of one turbine within a relative bigger grid. Therefore, it is necessary to determine the average reference wind speed for each wind farm in METRAS. The reference wind speed is chosen over all grid cells containing the wind farm itself. The area per grid cell covered by a rotor is defined as wind turbine mask.

Multiplying Eq. (3.2) by the wind turbine mask and adding this term to the basic momentum equation lead to the parameterization for wind turbines.

$$\vec{v} = \text{equation of motion} + \text{wind turbine mask} \times T' \qquad (3.3)$$

Due to the wind speed depending on thrust coefficient, wind turbines switch on and off autonomously, if wind speed is equal to the cut-in or cut-off velocity. Normally, the cut-in and cut-off wind velocities used in METRAS are 2.5–17 m/s (personal correspondence with M. Linde). Turbines and actuator disc are movable and are set orthogonal to the wind. The influence of the tower is ignored so far that no real obstacle is implemented. But to have an effect due to friction in case of less wind, pressure in the grid of the rotor will increase in case wind speed is lower than cut-in velocity. With these assumptions and the implementation of the ADC, several large wind farms can be represented in the model domain.

Wind turbines are simulated using an active rotor disc, and the changed wind field is parameterized adding a deficit term to the equation of motion.

In this study, the height of the tower counts 80 m, like the rotor diameter, which is a normal size of wind turbines in offshore wind parks.

A restriction is the different scales between the wind turbine and the grid box. Due to ratable consideration of wind change, the effect is a ratable projection on the whole grid box where the wind turbine is placed within. Although the turbine does not affect the whole box, a better solution is currently not available. This adaption leads in turn to an overestimation of the wake size, which has to be considered. As described in the SAR (synthetic aperture radar) data by the work of Li and Lehner (2012), wind farms often lead to one uniform wind wake, but under certain conditions, single wakes behind each wind turbine can be seen. Such single wind wakes cannot be considered here due to scales.

3.2   Data

Besides using METRAS data as forcing data, additional forcing data were necessary. These data are reanalysis data from ECMWF. An additional data set of ocean data comes from ship measurements, which are used to evaluate model results. Further details can be found in following sections.

3.2.1   Climatological and Reanalysis Data

Climatological and reanalysis data were necessary for the simulation of the North Sea and German Bight. METRAS uses ECMWF (European Centre for Medium-Range Weather Forecasts) data as meteorological forcing, as well as NOAA (*National Oceanic and Atmospheric Administration*) Optimum Interpolation Sea Surface temperatures for SST forcing.

HAMSOM uses Era-Interim data and a forcing mix of METRAS and ECMWF data for meteorological forcing. The mix of meteorological data was necessary because of the detriment caused by the METRAS model setup; only the atmosphere over the investigation area of the German Bight was simulated by METRAS. The atmosphere data over the remaining North Sea area are defined by another, the ECMWF, data set.

Atmospheric forcing includes 10-m wind fields, surface pressure, 2-m and 10-m temperature and humidity fields, precipitation, and cloud cover.

Horizontal resolutions of atmospheric fields are 1.5° for Era-Interim and 1° for ECMWF data, which cause a gridding to HAMSOM's horizontal resolution, and METRAS data have a resolution identical with that of HAMSOM.

Additional data necessary for ocean simulation are oceanic forcing at lateral open boundaries for ocean simulation, which consists of ocean temperature, salinity concentration, surface elevation, and river runoff. Thereby, those data are monthly climatological means based on the climatological ocean data set of World Ocean Atlas 2001 (WOA-01) by Boyer et al. (2005) for salinity and temperature. The river runoff is considered by runoff fluxes of 46 rivers along the North Sea coasts, gathered by Damm (1997) and O'Driscoll et al. (2012).

3.2.2 Measurements

The BSH supported hydrographical and hydrological measurements around the test wind park *alpha ventus*. The cruise aboard **VWFS** (Vermessungs-, Wracksuch-, und Forschungsschiff) WEGA allowed the use of Acoustic Doppler Current Profiler (ADCP) and CTD probe (Conductivity, Temperature, Depth probe) within 11th–13th of May 2013. The data set comprises three ADCP stations, which took measurements over 2 days and 39 + 3 CTD profiles within 1 day. Some impressions of the WEGA cruise are given in Appendix C.1. An explanation of the CTD probe and the ADCP instruments is documented in Chap. 9. Data presentation and evaluation are placed in Sect. 5.4.

3.3 Methodology

To analyze the effect of OWFs on the atmosphere and ocean, mainly model data were consulted. As mentioned, model results were based on METRAS simulations, which again were used as meteorological forcing for ocean simulation with model HAMSOM. Various simulations were carried out to investigate on the different factors triggering the OWF effect on the atmosphere and ocean. To fully capture the possible effects, the analysis is separated into two main approaches. Hence, the analysis is based on two types of simulation, shortened to **TOS**.

Type of Simulations
The first type of simulation, TOS-01, serves as an idealized approach to analyze physical aspects in theory, considering different factors with the aim of providing a common conclusion of the OWF effect on the atmosphere and ocean. The second simulation type, TOS-02, uses a more realistic approach, with simulations concentrating on the OWF effect on the area of the German Bight with the final overall aim of estimating possible impacts for life in that region.

The concepts of TOS-01 and TOS-02 are opposed to each other, as seen in Table 3.1. All simulations for each type of simulations were calculated twice, considering cases of nonoperating wind turbines (reference run REFr) and cases of operating wind turbines (OWFr). The difference between OWFr and REFr (OWFr–REFr) emphasizes the effect of the OWF on respective medium.

In the following, concepts of TOS-01 and TOS-02 are documented.

3.3.1   Model Box Simulations: TOS-01

Simulations of TOS-01 answer the purpose of the analysis of OWF's effect on the atmosphere and, especially, on the ocean under various external conditions like wind speed, size of wind farm, duration of OWF operation, depth of ocean, as well of computational issues regarding resolution and used forcing origin. TOS-01 also comprises the core of this study—a physical process analysis of the occurring dynamical changes in the ocean due to the OWF consisting of 12 wind turbines.

Model Area in TOS-01 (Model Box)
TOS-01 uses an idealized model area in the form of a box with the size of 240×240 km in the horizontal for both atmosphere and ocean models. The box is located in the German Bight, with the offshore wind farm *alpha ventus* in model center. Figure 3.2 shows the location and dimension of the model area. The offshore wind farm *alpha ventus* is located at 006.60° East and 54.00° North. For the ocean simulation, a model box (ocean box) is used. For METRAS, a similar box is used, but it is used for atmosphere simulation (atmosphere box). Therefore, the model area of HAMSOM and METRAS differs only by vertical resolution.

But the position of the model area on the map (Fig. 3.2) is important for HAMSOM because the ocean model uses an isogonic calculation, whereas METRAS is not isogonic but uses a flat area with grid sizes in kilometer unit. The isogonic approach allows a more precise application of Rossby radius due to the isogonic dependence of Coriolis parameter f. That difference between HAMSOM and METRAS in model design later asks for the interpolation and projection of METRAS data in the HAMSOM grid.

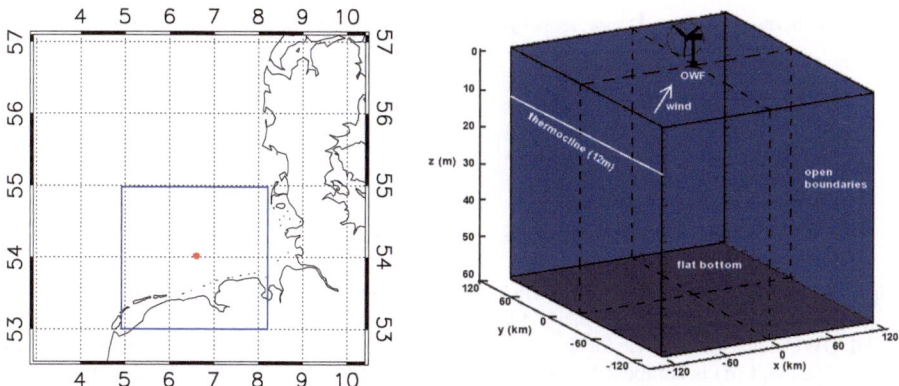

Fig. 3.2 Geographical location of the model area (*left*) and its arrangement as an idealized ocean box (*right*). The *blue square* within the North Sea marks the model area with the wind farm *alpha ventus* in the *middle*, whose position is given by a *red* point (*left*). The investigation area contains a box with a horizontal dimension of 240 km × 240 km and a maximal deepness of 60 m. Each grid box is sized 3 km × 3 km in the horizontal, which means a horizontal resolution of 2.5′ × 1.5′ in HAMSOM, and 2 m in the vertical. Topography, respectively land mass, within the box is neglected; the bottom is flat, and ocean boundaries are treated as open. The OWFs used are placed in the middle of the box. The atmosphere box in METRAS has the same horizontal dimension and resolution of 3 km × 3 km but differs, in comparison with HAMSOM, in the vertical resolution

Topography in TOS-01 (Model Box)

In both cases, atmosphere modeling and ocean modeling, the topography and bathymetry, respectively, are flat to elicit the sole effect of induced changes in dynamics by OWFs.

Offshore Wind Farms in TOS-01 (Model Box)

For analysis, different OWFs were used, which all were implemented around the center of the model area. The wind turbines are considered only in METRAS model due to the wind turbine parameterization, which is documented in Sect. 3.1.2. In ocean simulations, the OWFs are only implemented via meteorological forcing fields that were simulated in the METRAS model.

The size of the OWFs varies due to different experiments. One OWF, mostly used for analysis, consists of 12 wind turbines with an arrangement based on the German test wind park *alpha ventus*.

Additional tested OWFs consist of 48, 80, and 160 turbines arranged in four or rather eight rows; see Fig. 3.3. These different OWFs are used to evaluate changes on the atmospheric wind field and ocean due to different wind park sizes. The number of turbines is based on currently available and planned OWFs in the North Sea (approved wind farm projects within the North Sea listed under BSH 2013). Each wind turbine, implemented in METRAS, has a hub height of 80 m and a rotor diameter of 80 m. The rotator disc directly affects the vertical heights, ranging from

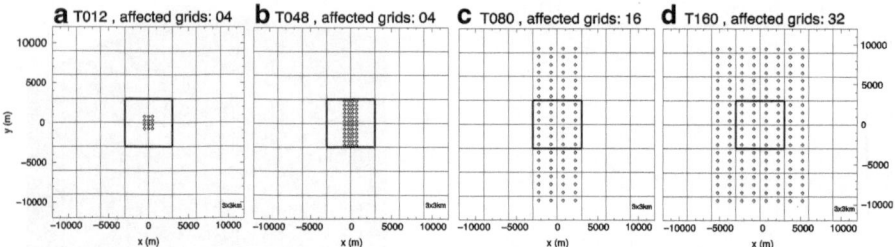

Fig. 3.3 Arrangement of wind turbines in TOS-01. Wind turbines are placed within grid boxes of 3 km × 3 km dimension and are marked by *red diamonds*. Box (**a**) shows the arrangement of a small OWF consisting of 12 turbines like *alpha ventus*, (**b**) consists of 48 turbines, (**c**) has 80 turbines, and (**d**) is composed of 160 turbines. The effect of each turbine counts ratable into corresponding grid boxes. Spaces between wind towers count 500 m in *x*- and *y*-directions in (**a**) and (**b**) and 1,500 m in *x*-direction and 1,000 m in *y*-direction in (**c**) and (**d**)

40 m up to 120 m. Although *alpha ventus* deals with bigger wind turbines, the smaller type of turbine is chosen based on the statistics of BWE.

Model Setup in TOS-01 (Model Box)
The model setup of the model box differs between simulations of the atmosphere and ocean. Thus, the model setup description is separated into METRAS and HAMSOM. The distinction is a result of an indirect coupling between the atmosphere and ocean by using METRAS data as meteorological forcing for HAMSOM and the assumption of idealized model conditions.

METRAS (Atmosphere Box Simulations)
As mentioned, METRAS simulations were done in collaboration with the Meteorological Institute of the University of Hamburg supported by M. Linde (Ph.D. candidate at MI-Hamburg since 2011). The aim of the simulations done was the creation of mainly a wind field including OWF information, which can be used as forcing for HAMSOM. Therefore, the METRAS simulation differs from the HAMSOM simulation, especially in time.

METRAS is used with a horizontal resolution of 3 km × 3 km. In the vertical, METRAS has an equidistant resolution of 20 m below 100 m, and above 100 m the vertical dimension of grid boxes is spread by a factor of 1.175. That vertical resolution is used for all simulations. The top is placed in 9,521 m.

As mentioned, the model bottom is flat due to the use of an ocean landscape type with a constant sea surface temperature (SST) of 15 °C over the whole simulation time. Additional variables prescribed for simulations are 'a' relative humidity of 70 % at 10 m, an one-dimensional start field that is chosen as stable dry, and geostrophic wind field. The forcing of METRAS only affects directly the upper layers above 1,000 m and only at the horizontal boundaries. Therefore, METRAS simulates the boundary layer by itself.

The geostrophic wind ug was prescribed with a velocity of ug = 8.0 m/s, respectively 5.0 and 16.0 m/s for analysis of the OWF's effect based on different wind speeds. The direction of the geostrophic wind is west and following the Ekman spiral; wind direction in 10-m height is southwest.

The whole simulation time for METRAS runs was counted 4 h. After 4 h, the simulated wind wake was defined as stable in METRAS. Thus, the last time step of METRAS result was used as forcing for HAMSOM.

Within atmosphere box simulations, METRAS runs comprise ten specified simulations. As mentioned, they differ in the amount of turbines and predetermined geostrophic wind. The specification of METRAS simulations, which produce forcing for HAMSOM, is listed and explained in Table 3.2. Simulations were done with operating wind turbines (OWFr) and without an OWF (REFr).

Additionally, a run including different durations of wind turbine operation was simulated; it means turbines were switched off and on. Operating wind turbines means the use of METRAS wind turbines parameterization. Details of the operation time are given in Chap. 4. That simulation is used to determine the evolution of the wind wake. To analyze the alternating OWF effect on the ocean, HAMSOM was forced with 10-min mean wind fields and not with a constant wind field over time, like in all other ocean simulations under TOS-01.

HAMSOM (Ocean Box Simulations)

Like in METRAS, the horizontal resolution of the ocean box in HAMSOM is approximately 3 km × 3 km, more precisely $2.5' \times 1.5'$. In the vertical, HAMSOM was run with two different ocean depths of 60 and 30 m. Vertical resolution is equally spaced from ocean surface to ocean bottom by 2 m. An ocean depth of 60 m is the maximal depth of Germany's EEZ where OWFs can be built. Also, 60 m is listed as a limit for OWF constructions [common statement of wind industry (Dena 2013)]. A depth of 30 m was implemented because the offshore wind farm *alpha ventus* was built in such a depth.

Time resolution counts 1 min. The simulation time comprises 5 days in sum, with the exception of one run that lasted over 30 days. Three days of each simulation are spent to spin-up the ocean and 2 days to establish the OWF's effect (Fig. 3.4).

The spin-up is requested to avoid wind-driven waves, Langmuir circulation, strong disruption at borders, and other implications initiated by a sudden strong wind forcing on a reposing ocean. During the 3 days, the wind speed increases from 0.0 to 6.0 m/s for the wind component u in x-direction and from 0.0 to 2.0 m/s for the wind component v in y-direction following a tangens hyperbolicus function. After 3 days, spin-up time for HAMSOM is forced by METRAS over 2 days every 10 min.

Boundaries of the ocean box are treated as open, which means that boundaries are defined according to the dynamical boundary equation.

Hydrographic starting conditions, of the temperature and salinity fields, are defined according to different possible conditions of the North Sea. Setup for ocean simulations under ocean box simulations comprises three different stratifications of ocean, TS01, TS02, and TS03. All three start fields of temperature and salinity are depicted in Fig. 3.5.

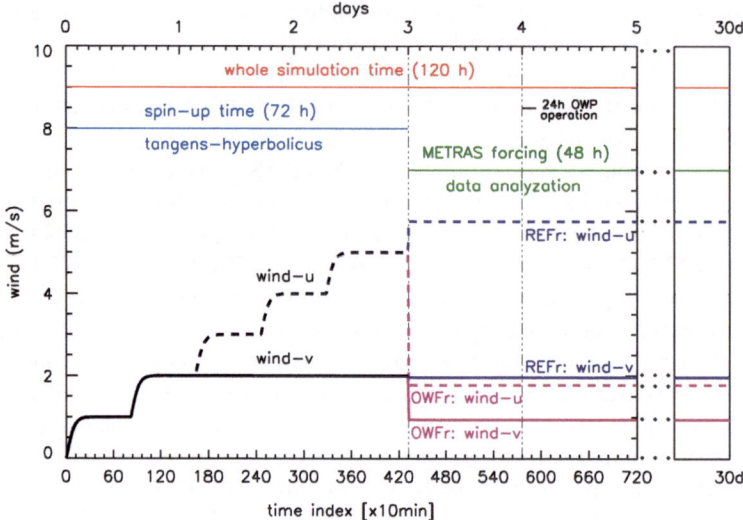

Fig. 3.4 Time setup of simulation under ocean box. The whole simulation time spans 5, respectively 30, days. Three days of spin-up is required. During spin-up, wind forcing is slowly increased to avoid unrequested dynamical effects within the ocean box. Over the last 2 days, a constant wind field of METRAS forces the ocean box. *Dashed lines* show wind speed of wind component *u*, *solid lines* of component *v*. Each component is separated into 'ref' (*violet*) and 'owf' (*pink*), which stand for different wind forcing fields regarding forcing without OWF signal and with OWF signal. The wind field forces the ocean every 10 min (time index of *x*-axis)

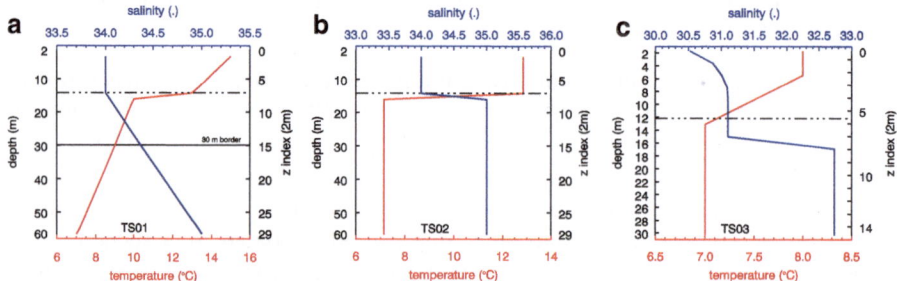

Fig. 3.5 Setup of temperature (*red*) and salinity (*blue*) start fields for whole ocean box. (**a**) For TS01, a linear decreasing/increasing of temperature/salinity is applied with values in accordance with North Sea conditions. (**b**) TS02 is a simplified stratification with only two layers being separated in 12-m depths. (**c**) TS03 is based on CTD measurements taken in May 2013 around the test wind farm *alpha ventus*

Stratification of TS01 is in accordance with the most common North Sea conditions, which mean warmer and fresher surface layer of 15.0 °C and 34.0 psu units and a thermocline in 12-m depths. Near-bottom temperature is set to 7.0 °C with a salinity of 35.0 psu.

The second *stratification setup, TS02*, only consists of two layers and is used to clearly define diffusion and exchange processes at thermocline. Values for the upper layer are 12.0 °C and a salinity of 34.0 psu; for the lower layer, it is 7.0 °C and salinity of 35.0 psu. Here, the thermocline is even placed in 12 m.

The *third stratification, TS03*, is based on CTD measurements around *alpha ventus* in May 2013 and is used to make the simulation comparable with measurements. In the case of TS03, the upper layer has a temperature of 8.0 °C and 30.5 psu and the lower one has 7.0 °C and salinity of 32.8 psu. The broad thermocline is located between 5 and 15 m.

Generally, HAMSOM is meteorologically forced with meteorological fields of 10-m winds, surface pressure, 2-m temperature, 2-m humidity, precipitation, and existence of clouds.

One aim of the analysis of ocean box simulations is the determination of the OWF's effect on the ocean due to the knowledge that wind turbines change dominantly the wind field. Therefore, in the analysis, wind forcing and atmosphere surface pressure were used only to detect the single effect of wind change. Two different wind forcing was employed, the one simulated by METRAS, including the wind farm's effect based on a disc rotor approach, and one considering OWFs by using an approach after Broström (2008), which is explained in Sect. 4.1. Broström's approach was applied on METRAS 10-m reference wind field without OWF influence, which was first converted to wind stress.

Full meteorological forcing was used based on METRAS data. But due to METRAS data availability, 10-m temperature and humidity fields were implemented instead of 2-m fields. An interpolation was not reasonable.

HAMSOM's ocean box is only meteorologically forced by METRAS data of last time step where the atmosphere is balanced and the wind wake is stable. The approach of using only one constant forcing wind field helps to define occurring processes in the ocean by OWF's wind wake. The forcing acts every 10 min on the ocean, which is calculated for each minute. Besides the last time step of METRAS, 10-min mean wind values are used for analyzing the effect of OWF operation.

In the end, HAMSOM is not forced by the variable wind but by wind stress. Therefore, wind values were transformed into wind stress by

$$\vec{\tau} = \mathrm{CD} \times \frac{\rho_{\mathrm{air}}}{\rho_{\mathrm{ref}}} \times \left|\vec{v}\right| \times 1.e3 \tag{3.4}$$

with $\mathrm{CD} = 0.0016$, $\rho_{\mathrm{air}} = 1.25$ kg/m^3, $\rho_{\mathrm{ref}} = 1{,}026$ kg/m^3, $\vec{v} :=$ wind vector.

Wind forcing incurs into the equation of motion as wind stress acting on the sea surface, but the turbulent shear stress is finally the actuating force for motion of the ocean. So the stress is a negative momentum flux into the ocean working as frictional force $\mathcal{F}$ at the sea surface:

$$(\mathcal{F}_x, \mathcal{F}_y) = \left(\frac{1}{\rho_{\text{ref}}} \frac{\partial \tau_x}{\partial z}, \frac{1}{\rho_{\text{ref}}} \frac{\partial \tau_y}{\partial z} \right) \tag{3.5}$$

Using those stratifications and spin-up of wind forcing, the actual effect on the ocean is simulated by a run using the baroclinic, prognostic mode of HAMSOM, Smagorinsky diffusion, and Lax–Wendroff advection scheme. That simulation is denoted as master simulation. Besides this main setup, various setups that differ a little bit from the main setup were considered. These different sensitivity runs were created to focus on theoretical details activating the occurring circulation pattern. The aim is to understand the physical principle standing behind the occurring ocean dynamic. Especially the analyses of horizontal and vertical momentum exchanges, as well as diffusion and advection processes, are important.

All these simulations are listed in Table 3.3, and runs considering OWF are shortly defined here:

- **Master Simulation *T012ug08 TS01HD60F01*:** the master simulation of TOS-01 includes forcing with wind field of 12-turbine OWF and ug = 8.0 m/s, temperature and salinity start field TS01, ocean depth of 60 m, and forcing comprises only one balanced METRAS wind and pressure field, which is kept constant over ocean simulation.
- **T012ug05 TS01HD60F01:** it is like the master simulation but ug = 5.0 m/s.
- **T012ug16 TS01HD60F01:** it is like the master simulation but ug = 16.0 m/s.
- **T048ug08 TS01HD60F01:** it is like the master simulation but with an OWF consisting of 48 turbines.
- **T080ug08 TS01HD60F01:** it is like the master simulation but with an OWF consisting of 80 turbines.
- **T160ug08 TS01HD60F01:** it is like the master simulation but with an OWF consisting of 160 turbines.
- **T012ug08 TS01HD60F02:** it is like the master simulation, but the wind forcing does not persist on one constant wind and pressure field; it includes different fields, as well as the effect of switching on and off of wind turbines.
- **T012ug08 TS01HD60F03:** it is like the master simulation, but full meteorological METRAS forcing is used.
- **T012ug08 TS01HD60F04:** it is like the master simulation, but forcing based on the Broström approach is used.
- **T012ug08 TS01HD60F01_BTM:** it is like the master simulation but with a manipulated HAMSOM source code to run simulation under barotropic mode.
- **T012ug08 TS02HD60F01_src*:** it is like the master simulation but with easier temperature and salinity stratification of only two layers (TS02) and manipulated HAMSOM source code to analyze exchange processes.

All simulations were computed twice, with a wind forcing neglecting OWFs—the reference run (REFr) and a wind forcing, including the signal of operating wind turbines (OWFr).

The results of meteorological forcing are presented in Chap. 4; the results of these runs for the ocean are represented in Chap. 5, as well as the source code manipulations of HAMSOM.

3.3.2 North Sea Simulations: TOS-02

The second type of simulation, TOS-02, is used for realistic runs of the North Sea and especially over the German Bight. Simulations of the North Sea are separated into two case studies aiming to estimate possible OWF impacts on the German Bight under more realistic conditions.

Case study I analyzes the OWF effect on the German Bight under a theoretical assumption of constant wind directions during a 1-day simulation.

Case study II is an adoption of a real passed meteorological situation of 16–19 June 2010.

The analyses of the OWF's impact on the German Bight consider a strong offshore wind farm demanded within the German EEZ. The OWF expansion for the German Bight that was used here is called '*B1-2030much*,' which consists of 8,590 wind turbines and is explained in details in Chap. 6.

Model Area in TOS-02
Figure 3.6 clarifies three model areas for simulations of the North Sea. The green encircled area comprehends the North European Shelf (HAMSOM NES); the red

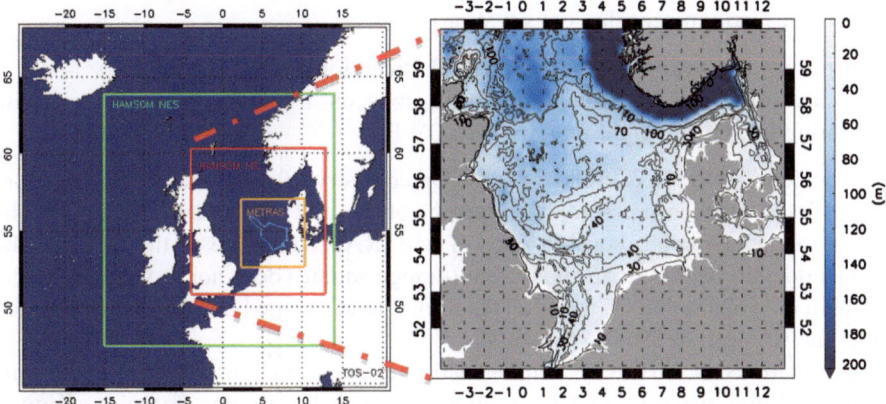

Fig. 3.6 *Left*: Overview of model areas in TOS-02. The *green area* marks the application to the European shelf (HAMSOM NES) of WOA-01 data used at boundaries. This model provides boundary data for the North Sea model (HAMSOM NS), shown by the *red area*. The *orange area* shows the model area of the atmosphere model METRAS, which fully includes the exclusive economic zone (*light blue line*) of Germany. *Right*: Bathymetry of HAMSOM model area NS in meters

encircled area defines the model area of the finally used HAMSOM North Sea (HAMSOM NS) simulation. The orange marked area shows the model area of METRAS, including Germany's EEZ (light blue).

HAMSOM NES (Green)

HAMSOM NES comprises ocean simulations of the North Sea and the north European shelf to provide initial data of surface elevation for ocean simulations over the smaller ocean domain HAMSOM NS. HAMSOM NES has a horizontal resolution of 20×20 km, and in the vertical the resolution counts 5 m from 0 to 50 m depths, 10 m from 50 to 100 m, 20 m between 100 and 200 m, and 50 m from 200 to 700 m. The topography of the ocean bottom is given in Fig. 3.6. The boundaries are treated as open in case of no coast. The temperature and salinity at the model boundaries are climatological data from WOA-01. HAMSOM NES is meteorologically forced with Era-Interim data every 6 h. The time step of simulations is 1 minute. Simulations were done for the years 2010 and 2011.

HAMSOM NS (Red)

HAMSOM NS comprises ocean simulations of the North Sea with a horizontal resolution of 3 km $\times$ 3 km. The vertical resolution is equal to the one of HAMSOM NES, and the topography is similar but logically spanned only over the model domain of HAMSOM NS. The boundaries are treated as open in case of no coast. The temperature and salinity at model boundaries are climatological data from WOA-01, and the initial surface elevation is based on calculations by HAMSOM NES. The time step of simulations over HAMSOM NS counts 1 min and provides a 10-min mean output.

Simulation time and meteorological forcing differs for simulations of case study I (analysis of OWF effect related to wind directions) and case study II (OWF effect under a realistic meteorological situation).

Setup for Case Study I: Constant Wind Directions

For case study I, 2×8 simulations were done on 1 June 2011: eight runs with operating OWFs and eight runs without OWFs.

The forcing for HAMSOM NS is adopted every 10 min and comprises 10-m wind fields, surface pressure, 10-m temperature, and 10-m humidity fields. The meteorological forcing consists only of METRAS data. Due to the fact that the model domain of METRAS in Fig. 3.6 (orange square) does not fit with the model domain of HAMSOM NS, the METRAS forcing at orange boundary was expanded over the whole HAMSOM NS domain.

Simulations of METRAS forcing data are based on a prescribed geostrophic wind of 8 m/s, meteorological conditions of an early summer day with an initial surface pressure of 1,000 hPa, 10-m temperatures of 17 °C and 10-m humidity of 70 %. The diurnal cycle of the sun is determined by 1 June 2011. METRAS uses a constant SST of 15 °C and a light stable and neutral moist atmosphere.

This setup was used for runs of eight different wind directions, which were assumed as constant over the entire simulation time of 1 day. The constant wind directions at the height of the geostrophic wind are

- N (000°),
- NE (045°),
- E (090°),
- SE (135°),
- S (180°),
- SW (225°),
- W (270°),
- NW (315°).

Based on Ekman spiral, the wind directions in 10 m height vary slightly from the geostrophic ones. Due to this idealized setup of METRAS, a combination with other (ECMWF) data was not reasonable. Disadvantages of that procedure are possibly physical abnormalities in the North Sea's circulation due to forcing expansion. However, such a side effect cannot be detected in the data set.

Setup for Case Study II: Real Meteorological Situation
For case study II, simulations of the German Bight were done with operating OWFs and without OWFs. The simulations, presented in this book, were done for the time range 16–19 June 2010. This time period is chosen based on meteorological situation and data availability for METRAS runs.

The full meteorological forcing is adopted hourly and includes forcing field of 10-m wind, surface pressure, 10-m temperature, 10-m humidity, total precipitation, and cloud presence. The forcing fields are a hybrid between METRAS data over the orange METRAS area in Fig. 3.6 and ECMWF data for the rest of the HAMSOM NS domain. At the boundaries of the orange METRAS area (Fig. 3.6), an interpolation was necessary. Simulations of METRAS forcing data were done from 16 to 19 June 2010 as well. As mentioned, the model area of METRAS simulations does not fit with HAMSOM NS; see Fig. 3.6. Over ocean, METRAS sea surface temperature is forced with the NOAA Optimum Interpolation Sea Surface Temperature. The meteorological forcing comprises 6 hourly fields of temperature, humidity, and the horizontal wind from ECMWF. The pressure field is then calculated by METRAS, which simulates its own dynamic.

The meteorological forcing field for simulations under HAMSOM NS needed some preparations. METRAS forcing data over ocean were interpolated and projected to HAMSOM grid, and ECMWF data were horizontally interpolated into HAMSOM grid. Also the use of a wind time step of 10 min and an hourly residual forcing asked for the need of time interpolation of 6 hourly ECMWF data. Although such strong timely interpolation is normally inappropriate, results show that only the orange METRAS area is affected by OWFs, and so areas affected by ECMWF interpolation do not influence the OWF effect analysis.

Simulations in TOS-02

As mentioned, simulations were done for two case studies. Simulations of case study I with constant wind direction are designated with T8590ug08wd*, while * stands for the wind direction in degree of 000°, 045°, 090°, 135°, 180°, 225°, 270°, and 315°.

The more realistic simulations of case study II are designated with T8590S01. Simulations of TOS-02 are listed in Table 3.3.

Table 3.1 Ocean simulation and analysis concepts at a glance

Main simulation	TOS-01: ocean box simulations	TOS-02: North Sea simulations
Description	Used for theoretical analysis Idealized ocean box	Used for analysis of German Bight Real North Sea area
Resolution	3 km × 3 km × 2 m	3 km × 3 km × z, with z-indices of 05 m (000–050 m) 10 m (050–100 m) 20 m (100–200 m) 50 m (200–700 m)
Topography	Flat bottom with depths of **30 and 60 m** no uplift due to bathymetry	North Sea's common topography (Fig. 3.6)
Time step (dt)	1 min	1 min
Wind time step	10 min	10 min, hourly
Output	10-min mean average	10-min mean average
Wind farm	One wind farm with different numbers of turbines (#**12,48,80,160**)	A couple of wind farms in the North Sea; total number of wind turbines is **8,590**
Wind forcing	**8 m/s** normal test run **5 &16 m/s** to analyze the effect of different wind speeds	**8 m/s** for run with constant wind direction; rest is variable and depends on meteorological forcing

Meteorological forcing	*Forcing variables (METRAS):*
	• 2 m/10 m of temperature
	• 2 m/10 m of humidity
	• Precipitation
	• Cloudiness
	• Wind stress
	• Wind speed
	• Surface pressure
	   – Mostly only wind (wind stress) and pressure forcing
	   – One simulation with full meteorological forcing included
		Forcing variables (METRAS & ECMWF):
		• 2 m/10 m of temperature
		• 2 m/10 m of humidity
		• Precipitation
		• Cloudiness
		• Wind stress
		• Wind speed
		• Surface pressure
		   – Full meteorological forcing
		   – Wind and pressure forcing only

Boundaries	Open boundaries	Open boundaries

Table 3.2 Overview of METRAS simulation

Type of simulation	Name	Abbreviation	Description
TOS-01 (Box Model)	M_T000ug05 M_T012ug05	**UG5**	REFr without wind turbines and ug = 5 m/s OWFr with 12 turbine OWF and ug = 5 m/s
	M_T012ug08 M_T000ug08	**UG8/T012**	REFr without wind turbines and ug = 8 m/s OWFr with 12 turbine OWF and ug = 8 m/s
	M_T012ug16 M_T000ug16	**UG16**	REFr without wind turbines and ug = 16 m/s OWFr with 12 turbine OWF and ug = 16 m/s
	M_T048ug08	**T048**	OWFr with 48 turbine OWF and ug = 8 m/s
	M_T080ug08	**T080**	OWFr with 80 turbine OWF and ug = 8 m/s
	M_T160ug08	**T160**	OWFr with 160 turbine OWF and ug = 8m/s
	M_T012ug08*onoff	–	OWFr with 12 turbine OWF and ug = 8 m/s and changing OWF operation
TOS-02 (North Sea)	M_T8590ug08**wd000** M_T0000ug08**wd000**	**N**	Constant wind direction N
	M_T8590ug08**wd045** M_T0000ug08**wd045**	**NE**	Constant wind direction NE
	M_T8590ug08**wd090** M_T0000ug08**wd090**	**E**	Constant wind direction E
	M_T8590ug08**wd135** M_T0000ug08**wd135**	**SE**	Constant wind direction SE
	M_T8590ug08**wd180** M_T0000ug08**wd180**	**S**	Constant wind direction S
	M_T8590ug08**wd225** M_T0000ug08**wd225**	**SW**	Constant wind direction SW
	M_T8590ug08**wd270** M_T0000ug08**wd270**	**W**	Constant wind direction W
	M_T8590ug08**wd315** M_T0000ug08**wd315**	**NW**	Constant wind direction NW
	M_T8590**S01** M_T0000**S01**	–	Meteorological situation of 16–19 June 2010

Main setup of METRAS under TOS-01 and TOS-02; see text. T000 designates no wind farm, and numbers greater than 0 behind T gives number of wind turbines in the simulation

Table 3.3 Overview of HAMSOM simulations meteorologically forced with METRAS for TOS-01 and TOS-02

Overview simulations		Description, explanation, characteristics, usage
TOS-01	**T012ug08 TS01HD60F01**	Master simulation
	T012**ug05** TS01HD60F01	Wind speed analysis
	T012**ug16** TS01HD60F01	Wind speed analysis
	T048ug08 TS01HD60F01	Wind farm analysis
	T080ug08 TS01HD60F01	Wind farm analysis
	T160ug08 TS01HD60F01	Wind farm analysis
	T012ug08 TS01HD60**F02**	Consistency analysis
	T012ug08 TS01HD60**F03**	Effect of met. forcing
	T012ug08 TS01HD60**F04**	Broström approach
	T012ug08 TS01**HD30**F01	Intensity in depth
	T012ug08 **TS03HD30**F01	WEGA adaption
TOS-01-src	T012ug08 TS01HD60F01_**BTM** T012ug08 **TS02**HD60F01_**src***	Physical analysis, different assumptions in source code
TOS-02	T8590ug08**wd000**	Effect by const. N wind
	T8590ug08**wd045**	Effect by const. NE wind
	T8590ug08**wd090**	Effect by const. E wind
	T8590ug08**wd135**	Effect by const. SE wind
	T8590ug08**wd180**	Effect by const. S wind
	T8590ug08**wd225**	Effect by const. SW wind
	T8590ug08**wd270**	Effect by const. W wind
	T8590ug08**wd315**	Effect by const. NW wind
	T8590**S01**	Effect based on real meteorological situation, June 2010

Main adjustments are based on number of wind turbines, meteorological forcing, and start fields. Each listed simulation here has got its pondo or counterpart without wind turbines
Abbreviations: **T** # turbine; **ug** geostrophic wind [m/s]; **TS** 01—common start field, 02—two-layer start field, 03—WEGA start field; **HD** HAMSOM ocean depth [m]; **F** METRAS forcing, 01—only wind, 02—full meteorological forcing, 03—on/off forcing, 04—Broström run; **BTM** barotropic mode; **src*** source code manipulation; **wd** wind direction [°]; **S01** summer season

References

Arakawa A, Lamb VR (1977) Computational design of the basic dynamical processes of the UCLA general circulation model. Methods Comput Phys 17:173

Backhaus J (1985) A three-dimensional model for the simulation of shelf sea dynamics. Ocean Dyn 38:165–187. doi:10.1007/BF02328975

Backhaus J, Hainbucher D (1987) A finite difference general circulation model for shelf seas and its application to low frequency variability on the North European shelf. Elsevier Oceanography Series. Elsevier, Amsterdam, pp 221–244

Becker G, Giese H, Isert K et al (1999) Mesoskalige Strukturen, Flüsse und Wassermassen-VariabilitÄt in der Deutschen Bucht, dargestellt durch die Kustos-Experimente und numerische Modelle. Deutsche Hydrographische Zeitschrift 51:155–179. doi:10.1007/BF02764173

Betz A (1926a) Windenergie und ihre Ausnutzung durch Windmühlen. Naturwissenschaften und Technik, Heft 2

Betz A (1926b) Wirbelschichten und ihre Bedeutung für die Strömungsvorgänge. Naturwissenschaften 14:1228–1233. doi:10.1007/BF01451780

Boyer T, Levitus S, Garcia H (2005) Objective analyses of annual, seasonal, and monthly temperature and salinity for the World Ocean on a 0.25° grid. Int J Climatol 25:931–945. doi:10.1002/joc.1173

Broström G (2008) On the influence of large wind farms on the upper ocean circulation. J Mar Syst 74:585–591

BSH (2013) Meeresnutzung-Windparks. In: www.bsh.de. http://www.bsh.de/de/Meeresnutzung/Wirtschaft/Windparks/. Accessed 20 Nov 2013

Carbajal N (1993) Modelling of the circulation in the Gulf of California. Reports Centre of Marine Climate Research 1–186

Damm P (1997) Die saisonale Salzgehalts- und Frischwasserverteilung in der Nordsee und ihre Bilanzierung. 259

Dena (2013) Offshore-Windenergie. In: www.effiziente-energiesysteme.de. http://www.effiziente-energiesysteme.de/themen/erneuerbare-energien/offshore-windenergie/. Accessed 1 Dec 2013

Hainbucher D, Backhaus J (1999) Circulation of the eastern North Atlantic and north-west European continental shelf – a hydrodynamic modelling study. Fish Oceanogr 8:1–12. doi:10.1046/j.1365-2419.1999.00009.x

Huang D, Su J, Backhaus J (1999) Modelling the seasonal thermal stratification and baroclinic circulation in the Bohai Sea. Continental Shelf Research 19:1485–1505

Li X, Lehner S (2012) Sea surface wind field retrieval from TerraSAR-X and its applications to coastal areas. In: IGARSS 2012–2012 I.E. international geoscience and remote sensing symposium, IEEE, pp 2059–2062

Linde M, Hoffmann P, Lenhart HJ, Schlünzen KH (2014) Influence of large offshore wind farms on urban climate; in preparation for the Meteorologische Zeitschrift

Mikkelsen R (2003) Actuator disc methods applied to wind turbines. Ph.D. Thesis, Technical University of Denmark

O'Driscoll K, Mayer B, Ilyina T, Pohlmann T (2012) Modelling the cycling of persistent organic pollutants (POPs) in the North Sea system: fluxes, loading, seasonality, trends. J Mar Syst 111–112:69–82

Pohlmann T (2006) A meso-scale model of the central and southern North Sea: consequences of an improved resolution. Continental Shelf Research 26:2367–2385

Chapter 4
Analysis 01: OWF Effect on the Atmosphere

The analysis of OWF's influence on the atmosphere and ocean is separated into three parts. Two parts study the effect of offshore wind farms on the atmosphere (this chapter) and on the ocean (Chap. 5) in theory by simulation type TOS-01 based on an idealized model area—the ocean box. Part 3 (Chap. 6) gives insight into the future of the German Bight regarding plans of wind farm development in 2030.

Before running the ocean model, it was necessary to find usable atmospheric forcing data, especially wind data considering wind turbines. Here, different appendages were chosen.

As mentioned before, wind turbines are used to transform wind energy into electrical energy. The main consequence of this transformation is a reduction of wind speed in the wind field behind wind turbines. This reduction of wind speed is called wake effect. The wake effect is the main component of this study, which drives and controls all following introduced processes and phenomena in the ocean. Hence, this chapter analyzes the wake effect with its incidents and conditions. The results are exemplified in observed and modeled wake effects.

4.1 Observed Effects

In the field of observed OWF effects on the atmosphere, there exists a handful of paper in literature. With a pioneer position in that field is M. Christiansen. She analyzes wind changes around OWFs using satellite data based on radar methods. An important example is Horns Rev, a Danish wind farm consisting of 80 wind turbines. Hasager et al. (2013) analyzed the occurrence of fog in the wake vortex behind turbines. The fog formation behind turbines uncovers the wind wake of each turbine. Hasager studied the appearance of fog, which appears due to advection of cold humid air over much warmer water surface; the possibility of upward mixing of saturated air from the surface into the cooler layer exist, and that can cause

© Springer International Publishing Switzerland 2015   35
E. Ludewig, *On the Effect of Offshore Wind Farms on the Atmosphere and Ocean
Dynamics*, Hamburg Studies on Maritime Affairs 31,
DOI 10.1007/978-3-319-08641-5_4

supersaturated mixture to develop and condense as fog or sea smoke. In contrast, cold-water advection fog occurs when warm moist air flows across colder water and the dew point temperature is reached such that fog forms (Hasager et al. 2013). Such impact of OWF can play an important role for local climates regarding moisture/cloudiness and temperature.

Christiansen identified the wind wake in evaluated radar images, which means the reduction of wind speed downstream of the wind farm (Christiansen 2006). While in the surrounding of Horns Rev the wind increases behind coastal area with strong shadowing effect of land, behind the wind farm a long plume of reduced wind speed was detected—the wind wake of the whole wind farm of more than 15-km lengths.

At this juncture, it becomes obvious that OWF effects on the wind field not only are a local phenomenon but also impact an area being much bigger than the OWF itself and have to be considered in models.

4.2 Modeled Effects

Modeling wind wake of wind turbines is a necessary tool for wind farm planning because a wind farm layout depends on the main wind direction and on the size of used rotor. To avoid turbulence from one turbine affecting another wind turbine, wind farms are designated with a minimum distance between individual turbines of around 7 rotor diameters (Jimenez et al. 2007; Meyers and Meneveau 2012) in the main wind direction. Wake modeling with high-resolution models like Large Eddy Simulations (LES) or Detached Eddy Simulation (DES) is widely in use. LES studies about wind farm wake are done by Jimenez et al. (2007, 2008), Wu and Porté-Agel (2010), Porté-Agel et al. (2011), and Hasager et al. (2013), just to name a handful of researches. Hasager, for example, did simulations with DES (Hasager et al. 2013). Such simulations provide a complex view of turbulences behind one or more turbines. Those details cannot be dissolved with mesoscale models like METRAS. However there exists a trend of implementing wind turbines into mesoscale models to evaluate OWF impact on the climate. The usage of mesoscale models with wind turbine implementation provides the advantage to analyze their impact on the weather and climate, but one has to deal with disadvantages regarding horizontal resolution.

The impact on the mesoscale is based on a theoretical approach after Broström (2008) and simulated results of the mesoscale model METRAS. Both approaches are adapted to HAMSOM later.

4.2.1 Mesoscale 01: Broström

Broström (2008) used a theoretical approach to analyze the influence of large wind farms on the upper ocean circulation by changing wind stress. After Broström, a wind stress in x-direction has the strongest disturbance in the y-direction, and so he defined the wind stress in two forms. The first is an assumption of a wind stress that is homogenous in the x-direction (Eq. 4.1), and the second is a more realistic one (Eq. 4.2) with a two-dimensional wind pattern (Broström 2008). This leads to the following formulation of wind stress:

$$\tau_x = \tau_{x0} - \Delta\tau_x e^{-\left(\frac{2y}{L}\right)^2} \tag{4.1}$$

$$\tau_x = \tau_{x0} - \Delta\tau_x e^{-\left(\frac{2y}{0.8L+0.2x}\right)^2} \max\left(e^{-\left(\frac{1-x}{L}\right)}\frac{x}{L}, 0\right) \tag{4.2}$$

with

$$\tau_{x0} := \text{windstress outside the influence of wind farm}$$
$$\Delta\tau_x := \text{change in the wind stress induced by wind farm}$$
$$L := \text{characteristic size of wind farm}$$

The advantage of this description is a cushy application. Here, a wind stress field with a mean wind stress of 0.012 N/m^2, which is based on reference wind speed of METRAS wind field simulation without wind farm operation and geostrophic wind speed of ug $= 8$ m/s, is supposed. Investigation area is based on simulation TOS-01 of 240×240 km with a wind farm of characteristic size of $L = 6$ km in the middle of this area. The changes of such wind stress field due to a wind farm are shown in Fig. 4.1 for two different wind directions, westerly and southwesterly. A westerly flow is even used by Broström, while here a southwesterly wind direction is added due to its frequent incidence in the German part of the North Sea (Loewe 2009). The reduction of wind stress by wind farm impact after Broström has an elliptical form with a maximum at wind farm's end of 0.0054 N/m^2 deficit, followed by a slightly wind stress increase in flow direction. Minimal values are 0.0064 N/m^2 for westerly flow and 0.0068 N/m^2 for southwesterly flow. Transverse to flow direction the wind farm changes the wind pattern in a symmetric way with the strongest deficit within OWF. The form of the wake is nearly independent of flow direction. Faint aberrations are attributed to the grid and so in flow direction into the front side of wind farm boxes and into the corner; see illustration at Table 4.1 under the line 'inflow of wind farm.' In sum, the wake shows a maximal reduction of 45–46 % with x-dimension of 39–42 km and y-dimension of 18–24 km. Here, x-dimension means the spread in flow direction and y-dimension, orthogonal to flow direction. The impact of the relative small wind farm of an area of 36 km^2 extends to an area of 702–1,008 km^2, so influencing a field, which is 20–28 times bigger than the OWF itself. For comparison, the city of Hamburg is measured around 755.3 km^2 (Haffmans 2005).

Broström's equations for wind stress changes give an estimation of how strong the impact of a wind farm can be. It is a quick test but a pure manipulation of wind stress, not based on physical principles; owing to these limitations, this method does not provide an optimal description of wind farm wakes. Considering this method,

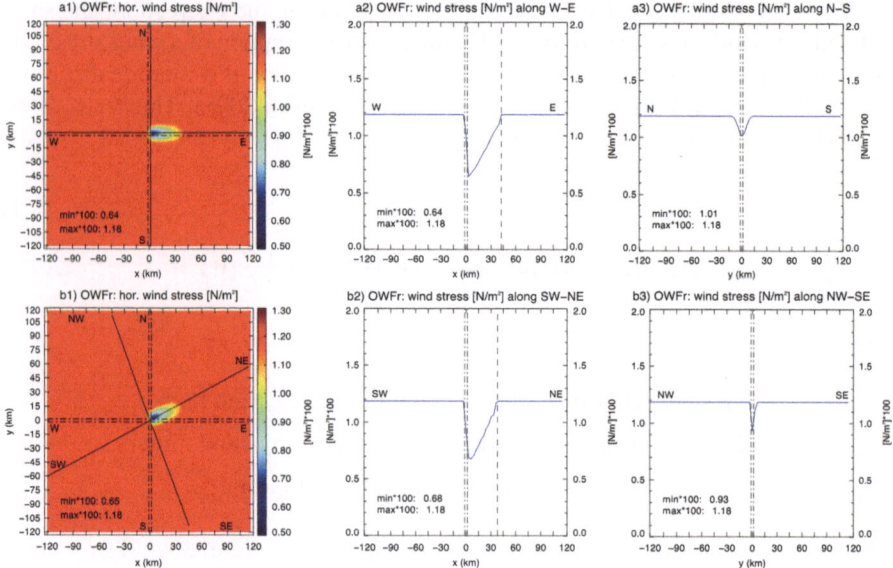

Fig. 4.1 Adopted Broström wind stress formulation on a reference wind stress field of 0.012 N/m^2 mean for two different wind stress direction and a characteristic wind farm length L of $L = 6$ km. Illustrations (**a1**)–(**a3**) show results for westerly direction of 270°. Illustrations (**b1**)–(**b3**) show results for southwesterly direction of 240°. Illustration (**a1**)/(**b1**) gives the horizontal wind stress field. OWF is placed in the middle, encased by *dashed dotted lines*, around $P(0,0)$. Horizontal resolution is 3 km × 3 km. *Solid lines* mark the cross sections W–E, N–S/SW–NE, NW–SE through the OWF. Illustration (**a2**)/(**b2**) represents wind stress along the cross section W–E/SE–NE and (**a3**)/(**b3**) along cross section N–S/NW–SE. *Dashed dotted lines* in the cross section plots mark the OWF's position; the *dashed line* in (**a2**), (**b2**) clarifies the wake dimension in wind direction. The wind stress along cross sections points to the wake behind and within the wind farm

Table 4.1 Key notes of Broström adoption

Broström (6 × 6 km^2 OWF, based on 10-m mean wind by ug $= 8$ m/s)	Max		Min	
Wind direction	W	SW	W	SW
Inflow of wind farm	➔■	⬈■	➔■	⬈■
Difference between OWFr–REFr [%]	No increase	No increase	−45.88	−44.92
$x =$ const: [N/m^2] × 100	1.18	1.18	0.64	0.67
$y =$ const: [N/m^2] × 100	1.18	1.18	1.01	0.93
Wake x-dimension [km]	42 (W–E)		38 (SW–NE)	
Wake y-dimension [km]	24 (N–S)		18 (NW–SE)	

only the impact of the wind field or, more explicitly, of the wind stress field in one special height can be described. The method cannot be easily adapted to other atmospheric parameters like temperature, humidity, and additional forcing fields normally needed for ocean simulations. Even characterized details of wind turbines are ignored, like rotor diameter or turbine power. Also, it will be problematic to adopt this approach to nonquadratic wind farm adjustments.

To account for such limitations and to describe such wakes more precisely, the wind wake is simulated using the mesoscale atmosphere model METRAS.

4.2.2   Mesoscale 02: METRAS

This section deals with the results of METRAS simulations of TOS-01 (atmosphere box) used for the theoretical analysis. Meteorological forcing for more realistic atmospheric situations, as North Sea simulations, is explained in Chap. 6.

METRAS advantage compared to, previously treated, Broström method is its physically based model frame and the employed wind farm parameterization, specified in Sect. 3.1.2. The wake is not estimated by an empirical formula but numerically solved. Therefore, METRAS' forcing is deemed to be the better alternative for simulations of the ocean, later due to an expected more realistic wake illustration. On one hand, the commonly used actuator disc approach allows a better definition of the form of the wake, as well as of its strength and dimension; on the other hand, specific details of wind turbines can be considered. Further, this way of wind forcing production provides data in the vertical and of all atmospheric parameters.

At first, the OWF's effect on the atmosphere as simulated in METRAS is analyzed, followed by an analysis on different wind speeds and wind farms, to estimate the wind wake based on different conditions and by an analysis of the OWF operation. In the following, changes between the run with wind turbines (OWFr) and the reference run without wind turbines (REFr) after 4 h of simulation are considered.

4.2.2.1   Analysis of OWF's Effect on the Atmosphere in METRAS

In general, atmospheric changes in METRAS are based on a 12-turbine wind farm and geostrophic wind with ug = 8 m/s (run M_T012ug08 in Table 3.2). Results for meteorological parameter pressure, wind, temperature, and humidity are illustrated in Fig. 4.2.

The main difference in the *wind field*, compared to Broström, is the form of the wind wake. The change of the wind field can be separated into three areas: a surge zone with weak decreased wind speed in front of the wind farm, the wind wake; a plume of reduced wind speed behind the wind farm; and two flanks of increased wind speeds flanking the wind wake. Here, the simulated wind wake is more than 120-km long with a width of 30 km and a maximum decrease of 4.42 m/s conforming to a decline of 71.65 %.

In the vertical, the zone directly affected by the rotor is 40–120 m. But the wind reduction occurs from the ground up to 250 m. In front of the wind farm exists a

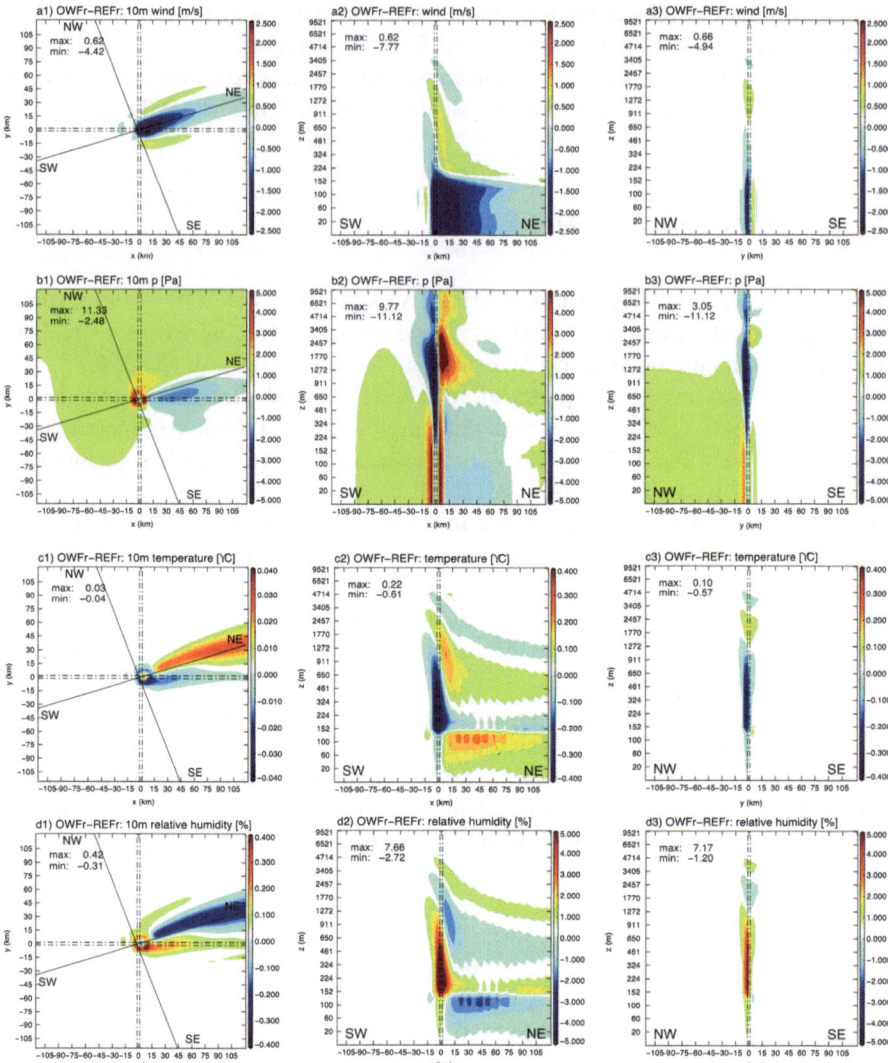

Fig. 4.2 Changes of meteorological parameter due to an OWF of 12 wind turbines based on a simulation setup of dry and stable atmosphere and a geostrophic wind of 8 m/s after 4 h of METRAS simulation. From top to bottom, the meteorological parameter are (**a1**)–(**a3**) wind speed (m/s), (**b1**)–(**b3**) pressure (Pa), (**c1**)–(**c3**) temperature (°C), and (**d1**)–(**d3**) relative humidity (%). Figures (**a1**)–(**d1**) show horizontal fields in 10-m heights; (**a2**)–(**d2**) show results along SW–NE cross section through OWF and (**a3**)–(**d3**), along the cross section NW–SE. *Dashed–dotted black lines* cross position of the OWF around $P(0,0)$ in the middle of the model area. *Solid lines* in (**a1**)–(**d1**) mark cross sections SW–NE and NW–SE through OWF. The hub height of wind turbines is 80 m; the rotor diameter is 80 m as well

downwind area reaching to 1,100-m heights. Behind the wind farm, above 250 m, wind speeds are increased by maximal 0.62 m/s.

Differences in the horizontal fields in 10-m heights of *pressure, temperature and relative humidity* are poor but show similar formations (Fig. 4.2).

The wind farm district, in the middle of the model area, is in 10-m heights 0.02 °C warmer and 0.2 % less moist due to operating wind turbines, and the pressure in the OWFr is reduced by 1.5 Pa. Outside of the OWF district occurs a circle with plume having opposite changes. There, the pressure is maximal 11.33 Pa higher in the run OWFr and temperatures are 0.04 °C lower and about 0.4 % moister. Within the wind wake district, 15 km northeasterly of the OWF area, temperature increases while moisture and pressure decrease. Near ground, the wind farm leads to very weak changes in temperature and humidity, while the pressure field changes at the surface at about ± 1 Pa. Strongest effects in wind direction of temperature occur around 230 m with a reduction of 0.61 °C, and the positive extreme of 0.22 °C is located around 130 m. The extreme values for changes in relative humidity are an increase of 7.66 % in 230 m and a decrease of 2.72 % in 130 m. The change of temperature and humidity is linked to up- and downwind in the vertical.

The warming of lower layers within the OWF and downstream of wind farm is connected with stability conditions and transport of potential temperature θ. In case of stable conditions, here ($\partial\theta/\partial z \gg 0$) for example, the vertical mixing due to OWF brings warm air down and cold air up, leading to a cooling above hub height, respectively at rotator disc, and a warming below. In case of unstable conditions ($\partial\theta/\partial z \ll 0$), induced turbulence would cause a mixing of cold air downward and warm air upward, producing a cooler surface. The similar process occurs for humidity. Dryer air is mixed down and moist air up, resulting in a frying below hub height and a moistening aloft. In this connection, an OWF also triggers surface fluxes. The ocean surface is colder than the above air, leading to a negative ocean–atmosphere thermal gradient, which again means a negative sensible heat flux. An increase of potential temperature due to OWF ends in a more negative sensible heat flux, and so more sensible heat is transferred from the atmosphere to the ground. The drying causes a positive ocean–atmosphere moisture gradient, which positively affects evapotranspiration. That context is also proved in Baidya Roy (2004).

The combination of changes in pressure and temperature over the ocean can favor cloud and fog formation. Here, any cloud and precipitation occur, perhaps due to stable atmospheric conditions, although having moist conditions with 68 % relative humidity.

In sum, the most important meteorological forcing parameters are the wind speed and pressure, which drive the upper ocean. Therefore, following analysis is concentrated only on changes in the wind field.

4.2.2.2 Analysis of Different Wind Speeds

A wind turbine must comply with several safety requirements defined by the International Electronic Commission (IEC). The IEC Technical Committee 88 prepares standards dealing with safety for wind turbine generator systems and produces standards for design and technical requirements. IEC defined four different turbine classes, which are determined by three parameters, the average wind speed at hub height, extreme 50-year gust, and turbulence (IEC 2005). These parameters, depending on location, will define the type of wind turbine generator (WTG) in connection with the size of wind turbine. Therefore, WTGs and rotors only operate in a limited range of wind speed depending on turbine class. Thus, a wind wake will be only produced at a certain wind speed range. And the wind speed is taken into account with respect to power by the cube. Thereby, stronger wind speeds result in higher power and a major energy transformation, which again results in a different strength of wind wakes behind wind farms.

To estimate the effect of wind speeds on the wind wake, three wind cases are analyzed. The simulation of these three wind cases differs by the input of geostrophic wind ug, which was set to ug = 5 m/s (run **UG5**), ug = 8 m/s (run **UG8**), and ug = 16 m/s (run **UG16**) (3.2). The wind farm consists again of 12 turbines being arranged over four grid boxes.

These results are illustrated in Fig. 4.3. The percentage changes of the horizontal wind field clearly show for all three wind speed cases the three zones of surge, wind wake, and flanked flanks. The wind wakes are larger than 120 km and exceed the model area, but the intensity of the wake is stronger with increasing wind speed.

Extreme values of minima occur within the wind farm. The stronger is the prescribed wind field, the stronger is the wake. The run of UG5 shows a reduction of 64 %, UG8 72 %, and UG16 77 %, compared to the reference run without OWF impact (Fig. 4.3a1–a3). The impact of the wind speed, respectively wind stress, on the OWF wake is nearly linear. Figure 4.4 pictures that relation between the prescribed geostrophic wind ug and the wake given by wind stress in N/m^2. But due to the small data set, a linear dependence cannot be generalized. Further simulations with wind speeds between ug = 8 m/s and ug = 16 m/s would be necessary for an approved statement.

In the case of UG16, the wake is less influenced by geostrophic force, compared to the other two wind speed cases, whose wakes are deflected more to the West, due to the stronger mean wind field (Fig. 4.3a1–a3).

The OWF-induced increase at wake's flanks does not follow the positive linkage. Here, the lowest increase of 9.01 % is given for UG8, the strongest with 9.77 % for UG5, which is close to the case of UG16 showing a wind speed rise of 9.65 %; see also Table 4.2.

Due to these flanks, the change in wind speed along the cross section NW–SE is not symmetric, as it is in case of the Broström method. Thus, the wake flanks do not only vary in intensity; their dimension even differs. While the area of wind reduction downstream the wind farm is quite constant in all three wind speed

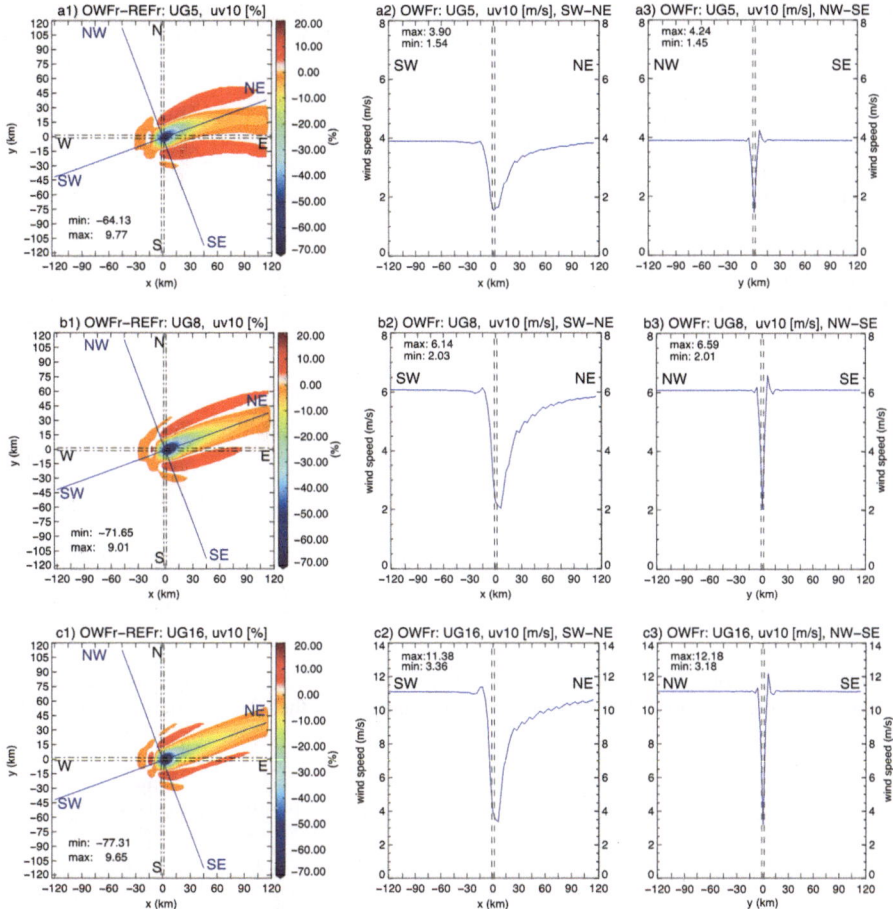

Fig. 4.3 OWF effect on wind field due to different prescribed wind speeds. (**a1**)–(**c1**) Percentage change of 10-m wind field due to 12-wind-turbine OWF. Wind wake reduction (**a2**)–(**c2**) in (SW–NE) and (**a3**)–(**c3**) orthogonal (NW–SE) to wind direction in case of three different ground wind speeds is presented after 4 h of METRAS simulation; values given in m/s. From top to bottom: results based on the different prescribed wind fields of ug = 5 m/s (**a1–a3**), ug = 8 m/s (**b1–b3**), and ug = 16 m/s (**c1–c3**)

cases, the wake flanks become shorter and thinner with increasing wind speed. A reason for this is that in case of higher wind speeds, occurring gradients are better balanced, which limits a strengthening of gradients. Therefore, it can be said that the area affected by OWFs decreases with wind speeds related to the area of the wake flanks.

Apart from that, a comparison of the three wind speed cases results in a small difference by maximal 10.0 % over the wind wake area and around 0.7 % over the wake flank zone. However, absolute differences results in greater discrepancies. How sensible the ocean reacts to these changes can be seen in Sect. 5.2.

Fig. 4.4 Dependence of prescribed geostrophic flow (ug in m/s) on wake magnitude (given in N/m² as wind stress being used as forcing for ocean simulations with HAMSOM) after 4 h of METRAS simulation. The relation is nearly linear, but due to only three points, it is not to be generalized

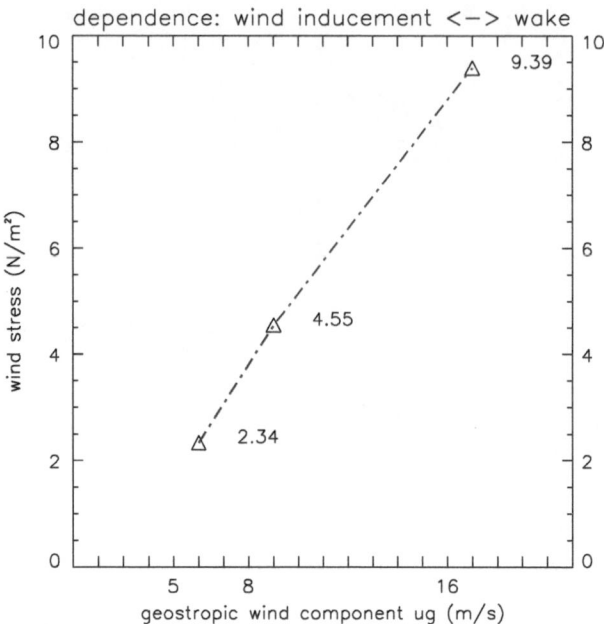

Table 4.2 Key notes of METRAS OWF simulation. OWF effect is separated into positive and negative changes over the 10-m horizontal (hor.) field and over the whole model area through the vertical layers (vert.)

METRAS (6x6km² OWF, 10m wind direction SW)		value horizontal and vertical	max. positive change	max. negative change
Turbine #	ug			
T012	8m/s	pressure (Pa), hor.10m	11.33	-2.48
		pressure (Pa), vert.	9.77	-11.12
		temperature (°C), hor. 10m	0.03	-0.04
		temperature (°C), vert.	0.22	-0.61
		humidity (%), hor. 10m	0.42	-0.31
		humidity (%), vert.	7.66	-2.72
		wind speed (m/s), hor. 10m	0.62	-4.42
		wind speed (m/s), vert.	0.66	-7.77
		wind speed (%), hor. 10m	9.01	-71.65
	5m/s	wind speed (%), hor. 10m	9.77	-64.13
	16m/s	wind speed (%), hor. 10m	9.65	-77.31
T048	8m/s	wind speed (%), hor. 10m	8.60	-71.65
T080		wind speed (%), hor. 10m	2.66	-68.18
T160		wind speed (%), hor. 10m	8.31	-80.05

On the basis of the cross sections, it is clearly seen that the magnitude of wake depends on the wind speed (Fig. 4.3b1–b3 and c1–c3). Another occurrence is the weaker the wind is, the stronger the wind speed within wake plume from NW to SE converges to zero and the wider the wake development from SW to NE is. That means along the section SE–NE until western model edge UG5 has a wake of 135 km, UG8 of 127 km, and UG16 of 125 km.

Besides wind speeds, also OWF conditions like the amount of wind turbines and the area being occupied can influence the magnitude of the OWF-induced wind wake.

4.2.2.3 Analysis of Different Wind Farms

Considering political aims, OWFs will be much greater than the used arrangement of 12 turbines within 4 grid boxes. This section shows how the wind wake will change due to different amounts of wind turbines, including 12 (**T012**), 48 (**T048**), 80 (**T080**), and 160 (**T160**) wind turbines (Table 4.2).

As mentioned, the wind farm constructions follow the rules of energy production. Due to the fact that each wind turbine produces a wake behind itself, the minimum recommended distances exist between them. Therefore, to be realistic, it is unfeasible to analyze different amounts of wind turbines within the four grid cells used before because an amount of 160 turbines cannot be placed in 4 grid boxes. That complicates a comparison of OWF wakes related to wind turbines. Nevertheless, keeping a quadratic form close to the middle of the model area elects the arrangement. Arrangement of wind turbines over the model area is shown in Fig. 4.5a1–a4. The prescribed geostrophic wind is ug = 8 m/s for all four wind turbine cases.

The 10-m horizontal wind fields for T012, T048, T080, and T160 are pictured in Fig. 4.5. The common change of the wind field is independent of the amount of wind turbines - in all cases a surge zone, the two flanks and the wake occurs. Logically linked to the occupied area by wind turbines, the dimension of the wind wakes varies. However, the range of changes in wind speed is close with a maximal reduction of 80 % for T160, 72 % for T012 and T048, and 68 % for T080; see Table 4.2. Spanning the wind turbines over more grid boxes intensifies the wake in magnitude and dimension.

Here, the wind turbines are distributed along y-direction, almost across wind direction, which defines the wake area.

Comparing OWFs of T012 and T048, they do not show obvious differences (Fig. 4.5), and these points to the fact that the distribution of wind turbines plays here a more important factor than the amount of turbines. A four times stronger wind farm, related to wind turbines, does not lead to a stronger wind wake. Here, the minima are the same (1.72 m/s) and the maxima slightly differ by only 0.02 m/s.

Figure 4.6 shows the relation between the amount of wind turbines, respectively OWF grid cells, and the wind speed based on overall extrema in 10-m heights. T012 and T048 have greater maxima compared to T160 and T080. T080 has a character of an outlayer. The wake in T080 is with 1.93 m/s the strongest, and the maximal wind speed of 6.24 m/s is the lowest. In this connection, the area in front of the wind farm T080 (between $x = -120$ km and $x = -30$ km) tends to the smallest mean wind speed of 5.76 m/s compared to T012, T048 and T160. In the case of T160, the mean wind speed in front of the OWF is 5.94 m/s and even smaller as in the case of

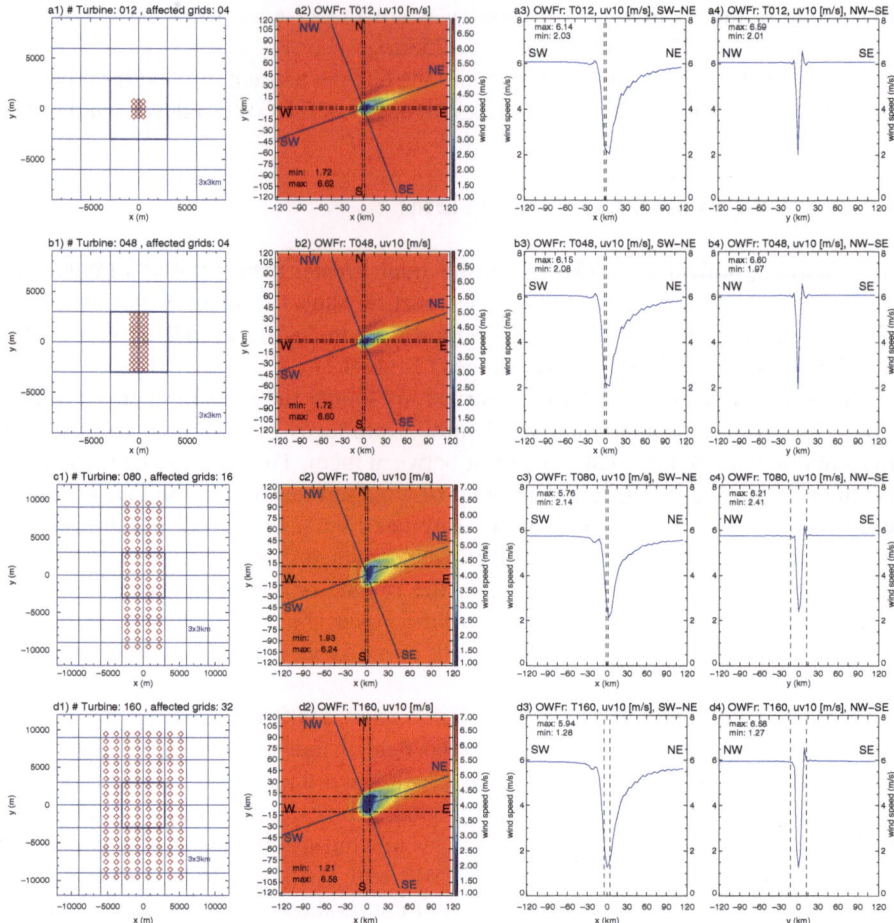

Fig. 4.5 OWF effect on the wind field due to different OWF districts and number of wind turbines. (a) Arrangement of wind turbines around the center of the model area; consider a zooming into the model area, and distances are given in *m*. *Red little diamonds* mark the position of wind turbines for (**a1**) 12 turbines, (**a2**) 48 turbines, (**a3**) 80 turbines, and (**a4**) 160 turbines. (b) 10-m horizontal real wind speed field of run OWFr after 4 h of METRAS simulation. *Horizontal black dashed–dotted lines* encase the OWF district around the model center. *Black solid lines* from SW to NE and NW to SE mark cross sections in and orthogonal to wind direction. (c) Shows the 10-m wind speed of OWFr along cross section SW–NE (**a3–d3**) and along NW–SE (**a4–d4**)

T012 and T048, but the effect is not as strong as in the case of T080. On one hand, such discrepancies between T080 and T160 lie in a weaker change in the pressure field around OWF for T080. On the other hand, the mean wind speed and the wake magnitude are obviously concentrated on the OWF district, and in the case of an OWF over more grid cells, which means in the case of a greater OWF, a better model accuracy is provided.

Fig. 4.6 Dependence of
number (#) of wind
turbines, respectively grid
boxes, on overall wind
speed maximum and
minimum of the 10-m wind
field after 4 h of METRAS
simulation time. Outliner
occurs with 16 affected grid
boxes

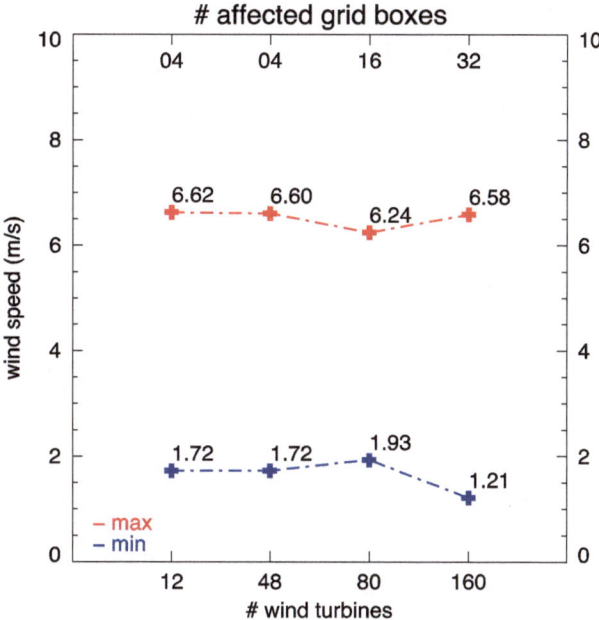

These sensitivity simulations show how difficult a nonmodeled parameterization for wind farms is, especially for wake flanks. There is not a clear linearity between affected grid cells and wind change, but the analysis of only three cases does not allow a final statement; for that, more cases would be necessary.

But an issue of scale becomes apparent because, as explained, the wind farm with 12 turbines results in nearly a similar outcome as that of a wind farm of 48 wind turbines. Only on the basis of wind speed, respectively wind stress, can one recognize a difference with a slightly expected stronger minimum in case of 48 turbines. This is explainable by the grid boxes covering the wind turbines. That side effect of modeling hampers a detailed statement of wind farm effects on the wind field. Hence, it is not exact to speak here of wind farms of 12, 48, or more wind turbines. A better way would be to specify wind changes based on the amount of turbines over an OWF district of, for example, four grid boxes. So in the case of 12 turbine wind farm used here, the effect is a matter of 12 turbines shared over 4 grid boxes affecting ratably an area of 6×6 km^2. The detected issue of horizontal resolution may overestimate the affected area and has to be considered.

However, a tendency of wake intensification with an amount of wind turbines, respectively the covered areas, can be supposed in reality. Hence, a bigger wind farm leads to a wider and stronger wind wake with a distinctive core of low wind speed within OWF district, and the structure of the influence will be mostly the same.

4.2.2.4   Analysis of the Consistency of METRAS OWF Effect on the Wind Field

Wind turbines only rotate in a limited window of wind speeds. In the case of METRAS, technical data of the wind turbine type NORDEX N80/2500 are considered in the wind turbine parameterization. Hence, the wind turbine parameterization only acts between 2.5 and 17.0 m/s (personal correspondence with M. Linde). This is leading to the question on how the wind wake changes in matters of different OWF operation modes. Thereby, METRAS provides another advantage, compared to using the Broström approach—the possible time analysis of wake development of a wind farm.

Figure 4.7 illustrates the 10-m horizontal wind field, simulated by METRAS (run M_T012ug08*onoff), for different time steps and OWF operation cases. The relevant step is, on one hand, the time when the OWF is switched off and the periods of switching on and switching off of the OWF. A nonoperating OWF in METRAS is still seen in the wind field because frictional resistance of rotor disc is considered. That leads to an increase within the OWF of 1 m/s. Therefore, a nonoperating OWF in METRAS is treated like a 'building,' which ends in an effect being comparable with a flow around a building.

Due to the dynamic pressure, the wind speed increases. In front of the OWF, the wind speed is reduced; pressure increases based on transformation of kinetic energy. At top and borders, a wind flow separation occurs with an increased flow due to depression. Behind the wind farm, a lag curl with a depression is expected; that is why even in the case of nonoperating wind farm, a wind wake is simulated behind the wind farm.

After turning on the wind turbines, which means using the rotor disc approach, the wind is suddenly reduced within the OWF district and at the flank the wind is increased due to depression (Fig. 4.7). With time, the wake grows and affects an area, which is significantly greater than the wind farm itself. With distance to the wind farm wind reduction slowly migrates to the wind speed of the surrounding. The produced wind wake and its flanks by OWF do not suddenly disappear after turning off the wind turbine operation; the main wind field advects wake and flanks.

With time, the effect of the OWF on the wind field can disappear by switching off the OWF. It can be said that in the ocean, the OWF effect is more dominant and is not erased within a few hours after turning off the OWF. However, it is important to understand and conceive the OWF effect on the ocean. These signals of the OWF on the wind field under different cases of OWF operation will be used as forcing for the ocean in the analysis of the OWF effect on the ocean.

To summarize, the OWF dominantly changes the wind field depending on wind speed and OWF conditions. The wind wake occurs within minutes of simulation time and becomes more intense in case of greater wind speeds and higher number of turbines. The herein presented 10-m wind fields were used as forcing for simulations of the OWF effect on the ocean, which is performed in Chap. 5.

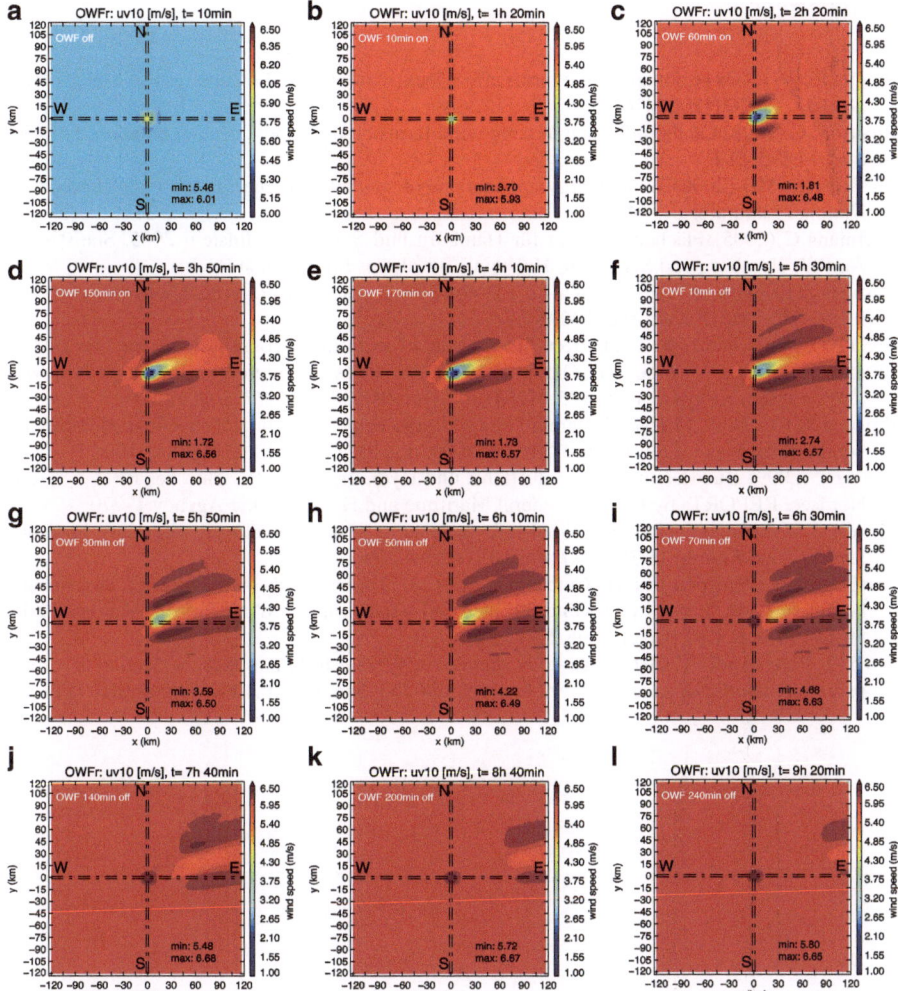

Fig. 4.7 Variations in the 10-m wind field due to different OWF operation mode of a 12-turbine OWF. Illustrations (**a**)–(**j**) show the development of the real (run OWFr) METRAS 10-m horizontal wind field for different time steps and OWF operation cases. The first wind farm is off (**a**) then is switched on (**b**), and turbines operate, which means METRAS wind turbine parameterization is considered (**c**–**e**). Later, OWF is switched off again (**g**). Turning off the OWF ends in advection of wind wake with the main wind field (**h**–**l**). Be aware of the different color bar in (**a**). The METRAS simulation times t are listed in the header of each picture

References

Baidya Roy S (2004) Can large wind farms affect local meteorology? J Geophys Res 109:D19101. doi:10.1029/2004JD004763

Broström G (2008) On the influence of large wind farms on the upper ocean circulation. J Marine Syst 74:585–591

Christiansen MB (2006) Wind energy applications of synthetic aperture radar. PhD Thesis, Risø National Laboratory 1–54

Haffmans C (2005) Flächenerhebung für Hamburg und Schleswig-Holstein 2005. Statistisches Amt für Hamburg und Schleswig-Holstein, Antsalt des öffentlichen Rechts

Hasager C, Rasmussen L, Peña A et al (2013) Wind farm wake: The horns rev photo case. Energies 6:696–716. doi:10.3390/en6020696

IEC (2005) info_iec61400-1{ed3.0}en. International Electronical Commission – Reports 1–9

Jimenez A, Crespo A, Migoya E, Garcia J (2007) Advances in large-eddy simulation of a wind turbine wake. J Phys Conf Ser 75:012041. doi:10.1088/1742-6596/75/1/012041

Jimenez A, Crespo A, Migoya E, Garcia J (2008) Large-eddy simulation of spectral coherence in a wind turbine wake. Environ Res Lett 3:015004. doi:10.1088/1748-9326/3/1/015004

Loewe P (2009) Bundesamt für Seeschifffahrt und Hydrographie – Berichte des BSH – System Nordsee. REPORTs by German Federal Maritime and Hydrographic Agency 1–270

Meyers J, Meneveau C (2012) Optimal turbine spacing in fully developed wind farm boundary layers. Wind Energy 15:305–317. doi:10.1002/we.469

Porté-Agel F, Wu Y-T, Lu H, Conzemius RJ (2011) Large-eddy simulation of atmospheric boundary layer flow through wind turbines and wind farms. J Wind Eng Ind Aerodynamics 99:154–168. doi:10.1016/j.jweia.2011.01.011

Wu Y-T, Porté-Agel F (2010) Large-eddy simulation of wind-turbine wakes: evaluation of turbine parametrisations. Boundary-Layer Meteorol 138:345–366. doi:10.1007/s10546-010-9569-x

Chapter 5
Analysis 02: OWF Effect on the Ocean

The possible effects of a wind farm on the atmosphere, especially on the atmospheric wind field, are summarized in the previous chapter. Based on the fact that wind farms will be heightened construct on the ocean, an analysis on how a changed wind field will affect the ocean's dynamics is of special interest in this context. At this juncture, different aspects are focused on.

Foremost, a common description of the impact on the ocean is given in subitem Sect. 5.1, introducing provoked changes in ocean dynamics and hydrography. Furthermore, a physical analysis explains and derives occurring changes in Sect. 5.2. Section 5.3 refers to different METRAS simulations for different wind speeds, amount of wind turbines, consistency of effect, and investigations regarding computational assumptions of ocean simulations, like forcing, depth of model area, and horizontal resolution, to fully capture all sides of the appearing phenomenon. Although HAMSOM is well verified over the years, the results are compared to measurements in Sect. 5.4 to estimate the dimension of OWFs' impact on the ocean and to evaluate the model data.

The main variables that are analyzed here are hydrodynamic and hydrological properties. Simulations are based on TOS-01 (box model, Sect. 3.3) with a mostly adopted geostrophic flow of 8 m/s; otherwise, then it is explicitly alluded to.

To point out the OWF effect on the ocean, the results are primarily presented by differences between a run with operating wind farm (OWFr) and a run without wind farm, the reference run (REFr). This dodge reflects the clear effect of OWFs because all changes in the dynamics do not occur in REFr.

5.1 Common Description of the Impact on the Ocean

The presentation of this section is based on the data of simulation *T012ug08 TS01HD60F01* to primarily detect OWFs' effect on the ocean. Here, it is assumed that the change of the wind field is the major effect of the wind farm on the ocean.

© Springer International Publishing Switzerland 2015   51
E. Ludewig, *On the Effect of Offshore Wind Farms on the Atmosphere and Ocean Dynamics*, Hamburg Studies on Maritime Affairs 31,
DOI 10.1007/978-3-319-08641-5_5

So only wind forcing is considered to analyze the natural effect of the wind wake on the ocean. Additional forcing besides wind forcing, like atmospheric temperature, is neglected.

It is known that the atmosphere strongly influences the surface upper ocean; especially, winds directly act with the ocean surface and play a key role on ocean flow (Wells 2012). Therefore, it is awaited that a change in wind field, here induced by a wind farm, influences the ocean system. In particular, since an OWF produces a wind wake of a dimension of several kilometers, as mentioned in Sect. 4.1, it is expected to consign a clear signal from atmosphere to ocean. First reflection on this context yields a reduction of upper ocean flow in wind wake region. But the effect of such wind wake on the ocean is more complex.

At the beginning of the result presentation, it is said that the wind's impact on the ocean is highly strong and the wake effect can be found in the ocean simulation after a few minutes of simulation time.

For a ready start into the subject matter, the OWF's effect on the ocean is first introduced by a moment analysis after 1 day of simulation, and then the effect evolution over 30 days is documented.

5.1.1   Moment Analysis of OWF Effect on Ocean

The moment analysis concentrates on the results of simulation *T012ug08 TS01HD60F01* after 24 h. Here, one constant METRAS wind field forces the ocean over the whole simulation time. Although the use of a constant wind forcing field over 1 day of ocean simulation is quite unlikely due to, for example, diurnal variations, that approach helps to clarify magnitudes of possible effects on the ocean due to the OWF's wind wake. The presentation of the OWF effect on the ocean after 1 day is chosen as a representative time step to illustrate possible impacts.

The velocity field at surface, which is, in a consistent manner, expected to react on wind changes, will be contemplated first. Knowing that wind's u-component dominates the form of wind change, this effect should be also identifiable at ocean flow's u-component at surface. Figures 5.1 and 5.2 present OWF effect on the ocean by horizontal velocity field, surface elevation ζ, and velocity components u, v, and w.

The *horizontal velocity field* in Fig. 5.1 indicates areas of reduced and increased flow around OWF. Averaged speed over the whole model area is 0.1 m/s, which is weak but thoroughly possible for a residual flow in the North Sea. The direction of flow is mostly southeast, made up of direction forced by wind and the Coriolis effect, which ends in a diversion of theoretical 45°. The change of the horizontal velocity does not agree with the OWF induced change of the horizontal wind field, as might be expected, because the velocity component v increases and acts as a result of occurring dynamics due to the OWF's wind wake. Nevertheless, the main

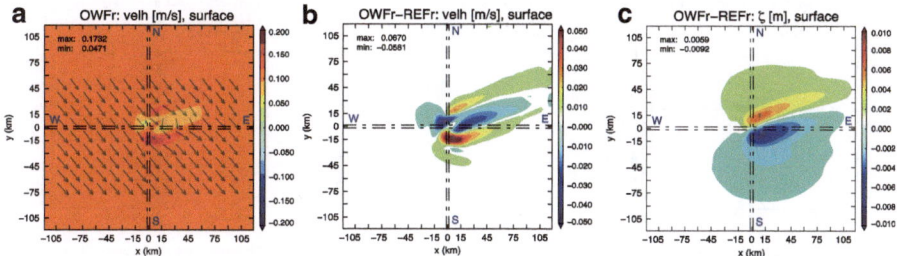

Fig. 5.1 12-turbine OWF effect on the ocean system at surface after 24 h of operating wind turbines. Shown variables are (**a**) the real horizontal velocity field (OWFr) and direction of flow and the OWF effect (OWFr–REFr) on (**b**) horizontal velocity field and (**c**) on surface elevation ζ. OWF is placed in the middle of the model area where *dashed–dotted lines* are crossing. The horizontal velocity field has a main wake behind the wind farm in wind direction, which is southwest. The velocity wake is flanked by an increase in velocities. Surface elevation ξ shows a dipole structure

decrease of flow is identified in the area of wind wake with -0.058 m/s after 1 day, but also regions of flow increase are produced with a maximal change of 0.067 m/s.

Surface elevation ζ shows a significant change (Fig. 5.1). Within the model area occur a maximum and a minimum of ζ, with division along wind wake axis in wind direction. That dipole structure of ζ has its increase north of the wind farm and wind wake, while its minimum is placed south of the OWF. The change induced by the OWF is in order of several millimeters; here, ζ counts an increase of 5.86×10^{-3} m and a decrease of 9.16×10^{-3} m, compared to the reference run after 1 day of simulation. Such formation of dipole is also postulated by Broström (2008) as one main effect that OWFs have on the ocean. Also, Paskyabi and Fer (2012), who adopt Broström's approach to analyze the response of the upper ocean on large wind farms in the presence of gravity waves, identified the disturbance in thickness of the upper ocean layer having the form of a dipole. The imbalance of the dipole's extrema is responsive later on but is correlated with additionally occurring dynamical effects, including circulations.

Having a look at each component of velocity, pictured in Fig. 5.2, the u-component clearly contains the wind wake information and is, like a fingerprint of the wind's u-component, respectively the wind field. The zone of wind wake, the raised flanks, and a distinct surge zone are clearly seen at surface in the horizontal. After 24 h, the wind wake affects the entire ocean strength of 60-m depths and reduces speeds up to 0.14 m/s. The increase of the u-component due to the wake flanks counts 0.08 m/s.

Especially in front of the wind farm and in the wake flank's areas, the v-component is reduced by 0.04 m/s, respectively increased by 0.05 m/s. Even here, the whole model depth of 60 m is affected, while changes of v-component due to the wind farm has its extremes around depth of the thermocline, in 12 m. Above this layer, the effect of the velocity component u counts stronger. The magnitude of the v-component depends on the wake flanks, as well as surface elevation and induced vertical motion.

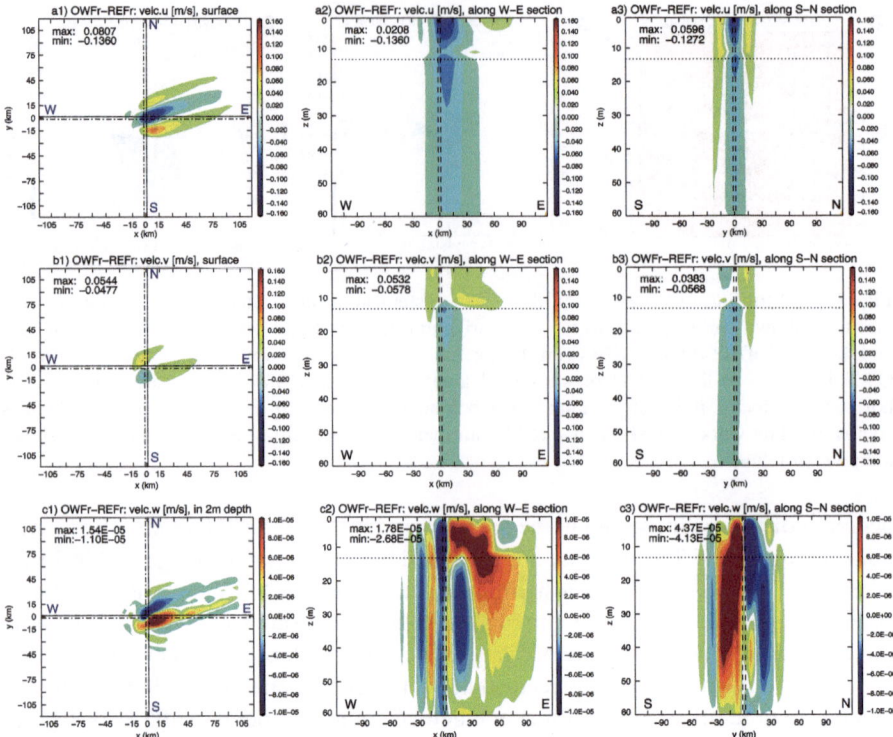

Fig. 5.2 12-turbine OWF effect on ocean velocity components after 24 h of operating wind turbines. The effect is presented for (**a1**)–(**a3**) u-component, (**b1**)–(**b3**) v-component, (**c1**)–(**c3**) w-component in the horizontal at surface (**a1**, **b1**) and in 2-m depths (**c1**) and along W–E (**a2**–**c2**) and S–N (**a3**–**c3**) cross sections through the OWF, which are marked with *solid lines* in the first domain. The dimension of the vertical component w is 3.78 m/d, which leads to an overturning after 15.89 days. Thermocline is defined in 12-m depths (*dotted line* in **a2/c3**–**c2/c3**) and the OWF is placed in the middle of the model domain around $x = y = 0$

Therefore, the most interesting velocity component is the vertical one. On the basis of the *vertical component w*, a triggered change in ocean dynamics due to the wind turbines is evident. In the horizontal at the sea surface, a wind farm provokes two main cells of positive and negative vertical velocity, spanning an area around the OWF of at least 10×10 grid cells, which means more than 900 km^2. These numbers underline a strong impact of wind changes induced within four grid cells, so an area of 36 km^2 where OWF is placed. The cross section through the model area from west to east and south to north of component w shows that vertical cells have been established within 24 h. In average, the cells have a size of around 30 km $\times$ 15 km and are affecting all ocean layers, especially the upper 30 m.

These upwelling and downwelling cells, with a speed of 0.04 mm/s, result in an overturning after 15–16 days. Even if that duration seems quite long, considering that in such time range the wind and the number of operating turbines can change, the induced vertical motion is an important phenomenon, which may also have an

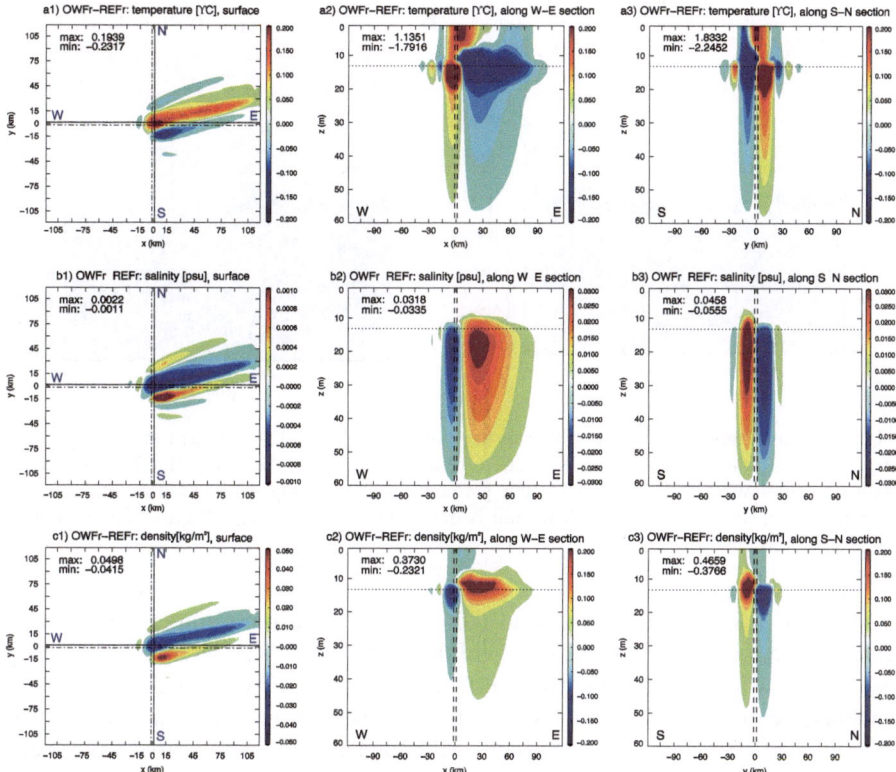

Fig. 5.3 12-turbine OWF effect (OWFr–REFr) on hydrographic conditions after 24 h operating wind turbines, be aware of different colorbars. The effect is presented for (**a1**)–(**a3**) temperature, (**b1–b3**) salinity, (**c1–c3**) density in the horizontal at surface (**a1–c1**) and along W–E (**a2–c2**) and S–N (**a3–c3**) cross sections through the OWF, which are marked with *solid lines* in the first domain. Thermocline is defined in 12-m depths (*dotted line* in **a2/a3–c2/c3**), and the OWF is placed in the middle of the model domain around $x = y = 0$

impact on the ecosystem. Beforehand, these vertical cells were durable, which will be discussed later under Sect. 5.3.

Besides hydrodynamic changes, the *hydrographic parameters* are affected too. Figure 5.3 summarizes results of *temperature, salinity, and density* after 24 h operating wind turbines, be aware of different colorbars. At surface, these variables depict the wind field.

Here, an increased sea surface temperature (SST) in the wake zone of 0.2 °C is recorded. The increase is adhered with a subsidence of surface elevation in the reference run, which results in little cooler reference SST compared to OWF run. The northern flank area is about 0.06 °C cooler, while in the southern flank area, where the main upwelling cell is located, a stronger decrease of 0.2 °C appears. Respectively, in those areas, the ocean becomes more/less salty with extreme changes of $1–2 \times 10^{-3}$ psu. Therefore, the wake area becomes less dense with

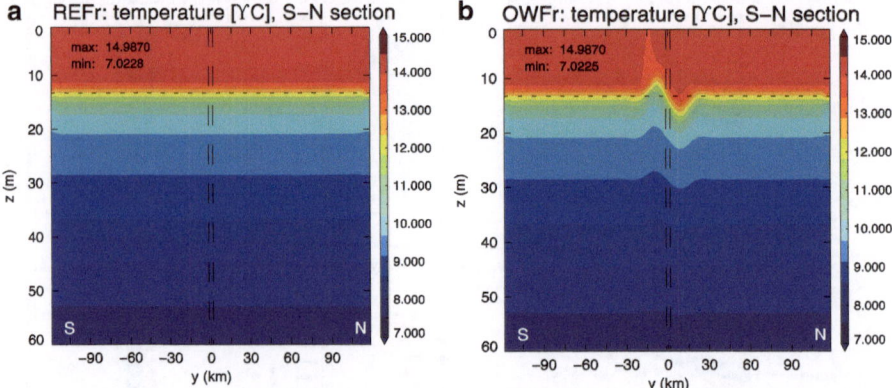

Fig. 5.4 Temperature stratification of run (**a**) without OWF (REFr) and (**b**) run with operating wind turbines (OWFr) along the cross section from S to N through the OWF after 24-h simulation with operating wind turbines. Values are given in °C. Operating OWF induces excursion of thermocline of about 10 m. That distortion is drawn through all layers. Thermocline is defined in 12-m depths

0.04 kg/m^3 and denser with around 0.05 kg/m^3 in the upwelling zone. Extreme changes are located around 12 m where the thermocline was set at the beginning, but still the whole ocean depth from surface to bottom is affected. In the vertical, the dependence on vertical motion is obvious because hydrographic changes occur in the region of up- and downwelling cells. Although along x-section from west to east vertical motion shows a more turbulent structure with several cells of opposite velocity directions, the change in the hydrographic fields are more homogeneous due to stronger vertical motion in the cross section from south to north.

Summarizing, an operating OWF induces a new oceanic dynamic around the OWF district. The important effect is the generation of up- and downwelling cells connected with changed hydrographic conditions, especially at the thermocline, compared to reference run. Operating OWF induces an excursion of thermocline of about 10 m (Fig. 5.4). This distortion affects all layers but weakens with depth having an exclusion of 4 m in 54-m depth.

The presented phenomenon of OWF on the ocean forms the subject of further examination in this study. Questions of analysis are the following: what exactly drive vertical motion, which processes control hydrographic conditions, how durable are those up- and downwelling cells, what magnitude is expected, and what conditions occur in reality?

5.1.2   Temporal Analysis of OWF Effect on the Ocean

So far, the theoretical effect of the OWF on the ocean after 24 h is investigated. With the help of temporal analysis, the first principle of the physics describing the phenomenon can be established. That time, analysis comprises a run of 1 month (30 days) with operating wind turbines. Again, here used assumption for ocean simulation is a constant wind field (last time step of METRAS run) forcing the ocean every 10 min by wind speed and direction for each time step. In reality, meteorological conditions will be never as constant as used here, but that proceeding allows an estimation of a possible OWF effect on the ocean by reaching an equilibrium ocean change. Also, the approach is used to avoid additional effects due to wind veering and gusts, for example. Such proceeding allows the best analysis of the development of the effect as well.

Changes in the *horizontal velocity field (VELH)* due to the wind refer to the wind forcing that incurs into the equation of motion as wind stress acting on the sea surface and pictured in Fig. 5.5a. The area of wind wake downstream of wind farm is projected on sea surface in the form of flow reduction from first time step (first time step is given after 10 min of simulation, including a 10-min mean) with a minimal speed of 0.01 m/s. With time, the wake flanks are even identifiable, resulting in an increase of horizontal velocity of 0.07 m/s and more, Fig. 5.5.

The direction of the horizontal velocity field at surface is veered by around 45°, compared to the wind direction due to friction and Coriolis force, so a southwesterly wind direction in 10-m heights leads to a nearly westerly (NWW) ocean flow in accordance with the Ekman theory. Although wind direction is constant with time, the direction of VELH varies from west to northwest close to OWF. That is connected with changes in the magnitude of velocity components, especially of the v-component. While during the first hours of operating wind turbines the horizontal velocity field looks like a fingerprint of the wind field, including a wake area, a surge zone, and flanks, the structure slightly changes with time. Hence, the u-component indicates the wind field, including an increase of magnitude with time. The change at velocity component is smaller, compared to the u-component. With time, the v-component shows a similar structure to the u-component, but intensified changes located around the OWF are stronger. Besides horizontal velocities, another indicator for a change in ocean dynamics is the change in surface elevation.

The reduction of the horizontal velocity field in the wake area affects the *surface elevation* ζ in a way that ζ increases easterly of OWF first, Fig. 5.5. Physically, the reduced ocean flow ends in a reduced transport of water masses in the wake area, which again results in a slack flow and so in an increase of surface elevation. Due to the law of conservation of mass, a counterreaction to that is recognized westerly of OWF; here, ζ decreases. So OWF leads to a dipole formation of surface elevation. The positive and negative cores of ζ changes move with time counterclockwise till the final dipole position is reached having an increase of surface elevation north of the wind farm and a decrease south of the wind farm. ζ changes are spread over the whole model area. A separation of the model area into increase and decrease of ζ is defined by the separation line $y = 0.25x - 30$ by setting a zero point within the OWF. The positions of ζ extrema at the beginning of the simulation (Fig. 5.5d1) are supported by the atmospheric pressure, presented in Sect. 4.2.2, and the velocity wake. The final distribution of the positive and negative ζ changes depend on the

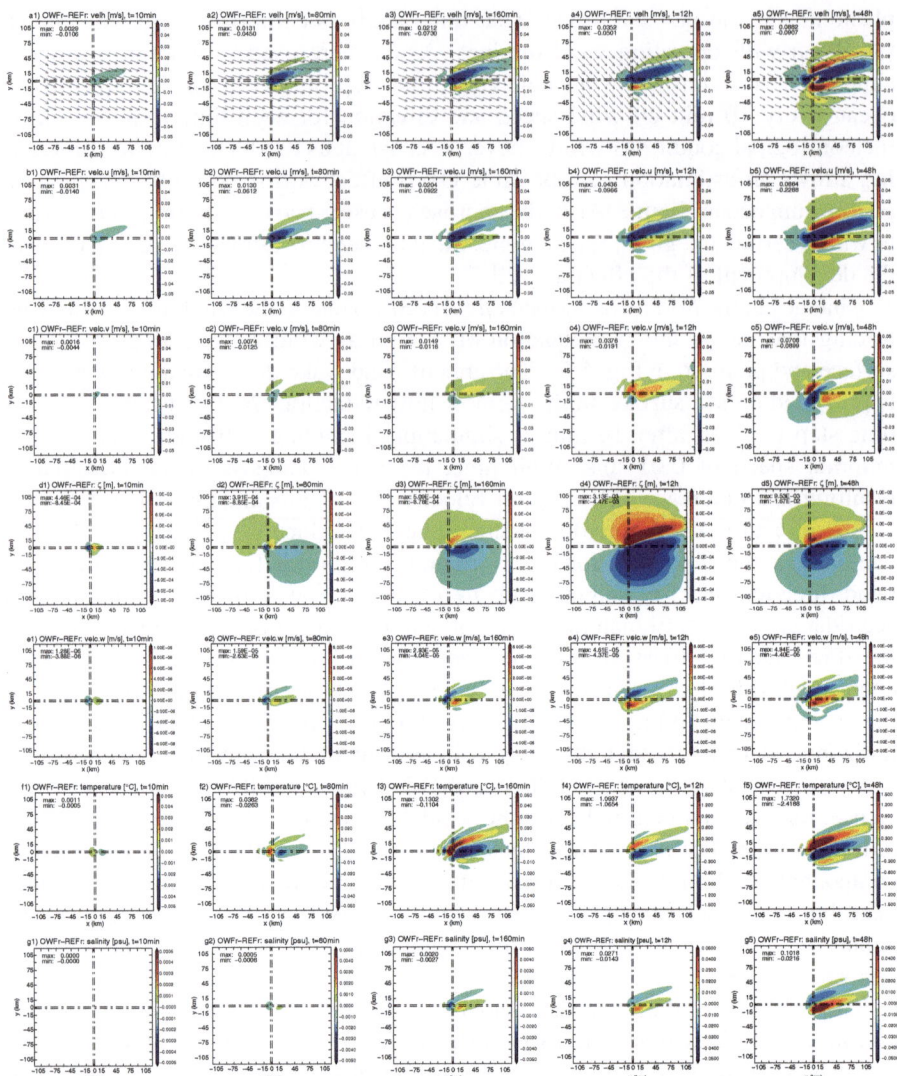

Fig. 5.5 12-turbine OWF effect (OWFr–REFr) on the ocean at five different time steps of ocean simulation at the surface for horizontal velocity field (**a1–a5**), velocity components u (**b1–b5**) and v (**c1–c5**), and surface elevation ζ (**d1–d5**) and at 12-m depths for velocity component w (**e1–e5**), temperature (**f1–f5**) and salinity (**g1–g5**). Results after 10 min (**a1–g1**), 80 min (**a2–g2**), 160 min (**a3–g3**), 12 h (**a4–g4**), and 48 h (**a5–g5**) of operating wind turbines are presented. The OWF is placed in the middle of the model area where *dashed–dotted lines* are crossing. Be aware of the different color bars at later time steps. *Black arrows* give direction of horizontal velocity field of run OWFr. Units of variables are listed in the header of each figure

increase of the velocity wake under geostrophic conditions and the Ekman transport, which is explained later in Sect. 5.2.3.

Connected with changes in the surface elevation, *vertical motion w* occurs following the same formation process around the OWF, Fig. 5.5. Two main cells of opposite vertical velocities increase by time showing downward speeds in the south of the OWF and an upward pointed *w*-component in the north. The reason for the up- and downwelling will be later examined, in Sect. 5.2. After 2 days, additional areas are affected by vertical motion besides two main cells. Here, especially, the area southwesterly of the OWF shows belts of vertical motion at depths of the thermocline.

Changes in *temperature and salinity*, representative of density, occur slower than changes in hydrodynamics, Fig. 5.5. Strongest changes are detected at the depth of the thermocline, while at the surface, effects are weaker. An increase/ decrease of temperature/salinity at the surface in the area of the wake is related to a change in surface elevation and wind forcing. The effect on the surface grows by time first, but the mostly warmer/less salty conditions are only a temporary effect. At the depth of the thermocline the OWF effect on temperature and salinity is caused by vertical velocities. Cooling and salinization dominate the model area downstream of the OWF after 1 day.

Although changes are variable within the first 48 h of simulation, formation of change is relatively constant after 2 days. Figures 5.6 and 5.7 illustrate changes during a time period of 30 days along the *y*-section as representative.

The *surface elevation* increases quite consistently with time, Fig. 5.6a. The analysis clarifies that maxima and minima stay stable after 27 days. The magnitude of negative tilt is stronger with a 33.22-mm change, while the positive change only counts 18.22 mm. The combination of horizontal and vertical compensation motions avoids a symmetric dipole.

The trend of *horizontal velocity field* and *components* is not as smooth as in the case of surface elevation. Over the first 2 days, a projection of wind field can be identified in the change of ocean flow, but within the OWF, the flow shows an increase from day three onward (see Fig. 5.6b). The velocity wake, formed in the beginning, is shifted more to the north and becomes weaker and horizontally thinner with time. Hereby, an additional second region of flow reduction occurs 20 km south of the OWF from day 10 on. Therefore, the area of a southerly wake flank spreads more to the south. Maximum changes in the flow are 0.18 m/s increase and 0.09 m/s decrease at the beginning and 0.08 m/s from day 14 on. The upcoming second flow wake is a result of an intensified positive *v*-component south of the OWF (Fig. 5.6d), as reaction to changed surface elevation. Within the OWF, the *v*-component is more negative in the run OWFr than in REFr, while the *u*-component is positive in both cases, with the exception of the OWF and wake area in OWFr. That leads to a higher horizontal velocity within the OWF area in the case of OWFr, compared to REFr. However, the *u*-component and *v*-component reach their constant level earlier than the surface elevation. After 5 days, they have within the OWF a reduction of −0.32 m/s for *u*-component and −0.12 m/s for *v*-component

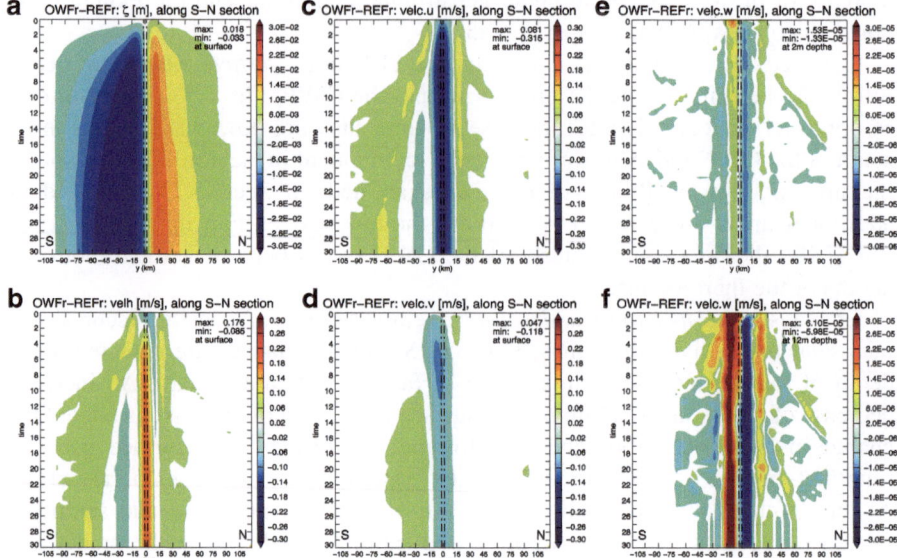

Fig. 5.6 Development of 12-turbine OWF effect (OWFr–REFr) with time along S–N cross section through the OWF at surface for (**a**) surface elevation ζ, (**b**) horizontal velocity field, (**c**) velocity component u, (**d**) velocity component v (**e**) at 2-m depths and (**f**) at 12-m depths for velocity component w. Time range comprises 30 days of constant operating of wind turbines. OWF district is marked with *dashed–dotted lines*

(see Fig. 5.6c–d). A maximum increase counts 0.08 and 0.05 m/s for u- and v-components.

The magnitude of the *vertical velocity* cells increases with time and is shown in Fig. 5.6e–f. The dimension of the cells leads to a diameter of 15 km along the y-section for both. At the surface, their increase stops and the magnitude of cells pulses a little bit due to the horizontal velocity field. In 12-m depths at the thermocline, the cells become nearly symmetric with maximal velocities of around 6.0×10^{-5} m/s, which is in accordance with 5.18 m/d, which again would end in an overturning after 11.57 days. Besides the two main cells, additional areas are affected by mostly downwelling from day 6 on.

While vertical cells appear to be symmetric after 20 days of simulation, changes in the *hydrographic conditions* are dominated by the cooling of the southern area at the surface, which becomes saltier and so denser, Fig. 5.7. A cooling is expected due to the transportation of warmer water to the depth and of cooler water to the top. At the thermocline, the warming is more distinct and the decrease with time is clearly visible, even its dispersion in the horizontal. The warming in the south is connected with an additional vertical downward motion between days 9 and 20. At the surface, a maximal decrease of $-1.8\ °C$ is simulated and at the thermocline, of $-2.95\ °C$, and the warming starts at the thermocline with 1.75 °C, Fig. 5.7. While temperature/salinity maxima are first concentrated at the vertical motion cells,

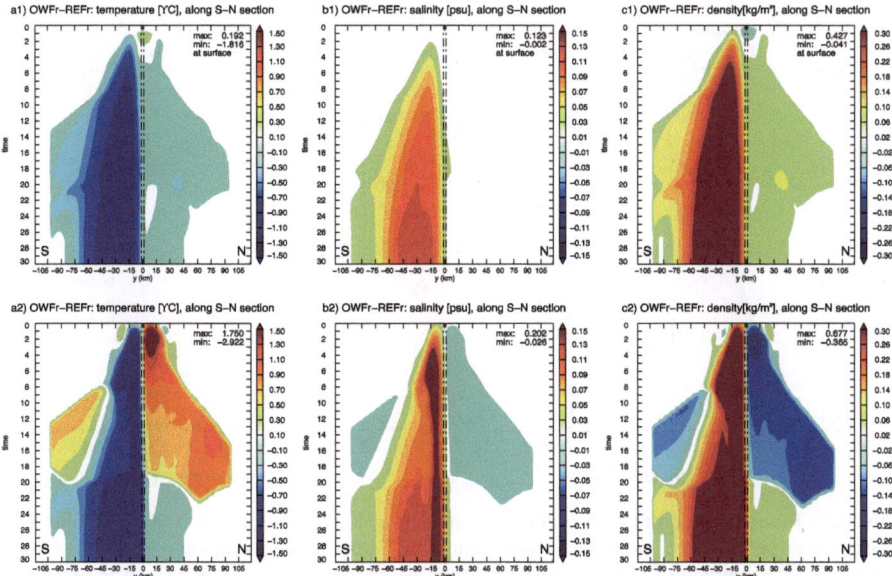

Fig. 5.7 Development of 12–turbine OWF effect (OWFr–REFr) with time along S–N cross section through the OWF at surface (**a1–c1**) and at 12-m depths (**a2–c2**) for temperature (**a1–a2**), salinity (**b1–b2**), and density (**c1–c2**). Time range comprises 30 days of constant operating of wind turbines. OWF district is marked with *dashed–dotted lines*

horizontal processes seem to support exchange by time over the model area, which first affects the surface.

If different time steps are compared at the beginning of the simulation, it is apparent that the upwelling cell has faster vertical velocities than the downwelling cell. Due to that, cooling is slightly faster than warming. Therefore, it also takes longer to warm lower ocean layers.

Figure 5.8 shows temperature profiles for different time steps of the ocean simulation at two points 6 km south and north to the OWF center along the S–N cross section. In the case of downwelling, the thermocline drops from 12 to 20-m depths. In the area of upwelling, the thermocline rises by more than 10. The change in thermocline depth grows with time; its magnitude depends on location.

Due to the stable distribution at the beginning, the thermocline cannot be fully eroded, but in the case of a more realistic temperature distribution, the ocean would be stronger mixed, which is shown in Sect. 5.4. Compared to the beginning of the simulation, the temperature at the upper layer undergoes a decrease of an average of 1.5 °C within 30 days southerly of OWF and leads to a temperature gradient between south and north of approximately 2 °C.

Details of physical processes linking and triggering the OWF induced phenomenon on the ocean system are analyzed in the next sections.

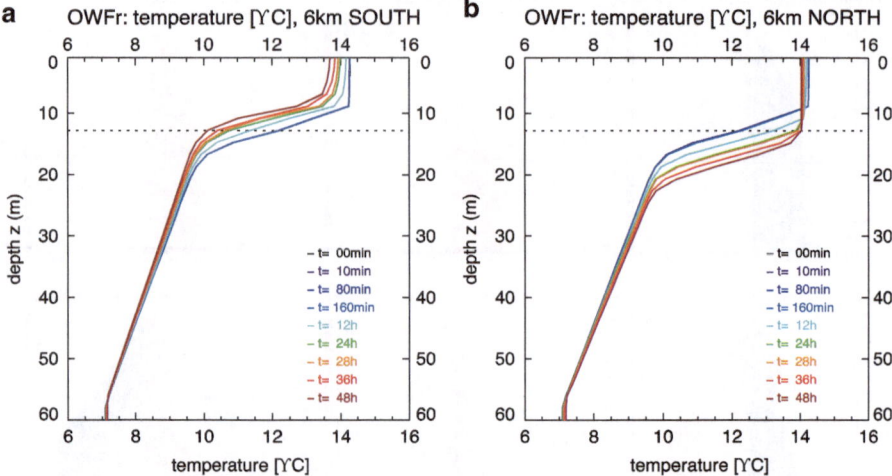

Fig. 5.8 Temperature profiles of run OWFr considering 12 wind turbines for different time steps *t* at two representative positions along S–N section. (**a**) Temperature profiles at 6 km south to OWF center. (**b**) Temperature profiles 6 km north to OWF center. Hence, (**a**) is placed in upwelling region and (**b**) in downwelling region. Thermocline is pushed with time up and down depending on signs of vertical velocity component *w*. The different time steps of OWF operation are 0 min, 10 min, 80 min, 160 min, 12 h, 24 h, 28 h, 36 h, 48 h, starting with the coloring from *black* to *red*

5.2 Theoretical Analysis of Rising OWF Effect on the Ocean

The effect of an offshore wind farm on the ocean includes modifications in ocean dynamics and hydrography, as documented in Sect. 5.1. The cause of vertical mixing, indicated as upwelling and downwelling, is associated with changes of the hydrographic components temperature, salinity, and density. An induced change of the ocean's stratification is of special relevance. Therefore, this section aims to understand the impact on the ocean due to operating wind turbines by analyzing the physical conditions, especially influencing vertical processes, step by step with the help of a sensitivity study.

The preparation of the sensitivity study accounts for the following facts and orientations:

i) **An OWF results in up- and downwelling cells around the OWF district.**
   In common, local mixing of the upper ocean is predominantly forced from the state of the atmosphere directly above it (Moum and Smyth 2001). Main processes for mixing are convection, wind forces, precipitation, and ice on the upper ocean (Moum and Smyth 2001). Here, the variable wind can be identified as the main impulse of the OWF effect on the ocean so far. The wind is the most important atmosphere component of this analysis because an intensified vertical motion occurs by using a forcing covering wind as meteorological forcing only.

Additionally, topographic irregularities can be excluded as a mechanism for incipient vertical velocities because of using a flat ground in simulations. But what exactly triggers changes in the vertical?

On the basis of the OWF induced wind change and related changes of surface elevation and considering time analysis, we can assume that vertical motion is driven by these horizontal changes at the surface, which affects the barotropic pressure field. To review, the importance of stratification under barotropic conditions is presented in Sect. 5.2.1.

ii). **An OWF results in an increase and decrease of hydrographic values (around the thermocline).**

The main processes supporting changes in the hydrography on short time scales are diffusion and advection of temperature, respectively salinity. The question is now whether horizontal gradients dominantly impact the OWF effect on hydrography or whether vertical advection mainly causes such changes. Hence, the impact factor of diffusion and advection and the exchange of momentum in the vertical and in the horizontal are analyzed in Sect. 5.2.2.

Sensitivity Analysis

On these grounds, the physical analysis comprises various sensitivity runs, which are listed in Table 5.1. All runs underlie the main setup of TOS-01 (Box-Model) with a 12-turbine wind farm over grid cells in the middle of the model domain and a prescribed geostrophic wind of ug $=$ 8 m/s. To analyze the physics of the emerging phenomenon, HAMSOM's equation within the source code (src) was prepared so far to receive the impact of different physical processes.

The simulation name for simulation under barotropic conditions is *T012ug08 TS01HD60F01_BTM (BTM)*. The corresponding effect-reference simulation is *T012ug08 TS01HD60F01 (BC)*. BC is the master simulation presented in the previous chapter considering the full HAMSOM code with 3D baroclinic primitive equations and an undistorted diffusion and advection scheme.

Simulations considering vertical and horizontal exchanges are designated as *T012ug08 TS02HD60F01_src** ('*' stands for modification listed in Table 5.1) and are explained in context. The corresponding effect-reference simulation is *T012ug08 TS02HD60F01*. That effect-reference run differs from master run BC by temperature and salinity stratification. Here, temperature and salinity start field TS02 (Fig. 3.5) is used. That TS stratification consists only of two layers. The reason for another TS description is the fact that the vertical exchange processes affect layers above and around the thermocline. Those layers can become indistinct due to the fact that such processes leads to a vertical mixing and so to a change on the conditions of the TS stratification. Hence, the impact of each process at the thermocline is not clearly identifiable. Therefore, the temperature and salinity field was simplified to two main layers, which allows a distinctive detection and analyses of the influence of exchange processes via and around the thermocline.

All setups listed in Table 5.1 were simulated in the case of OWF and in the case of no OWF to picture the sole effect after 1 day of the simulation.

Table 5.1 Overview of runs for the sensitivity study, commonly based on TOS-01 (Box-Model)

Main run	Description
i) Analysis of Barotropic Cause [temperature and salinity start field: TS01]	
T000ug08 TS01HD60F01 **T012ug08 TS01HD60F01**	Full HAMSOM model code with 3d baroclinic primitive equations and diffusion, advection scheme, called *master run*
T000ug08 TS01HD60F01_BTM T012ug08 TS01HD60F01_BTM	Master simulation but no baroclinic pressure component, TS changes barotrop, *abbreviation BTM*
ii) Analyses of Barotropic Cause [temperature and salinity start field: TS02]	
T000ug08 TS02HD60F01_src50 **T012ug08 TS02HD60F01_src50**	Like master simulation but with TS02, no exchange limitation, called *normal run*
T000ug08 TS02HD60F01_**src51** T012ug08 TS02HD60F01_**src51**	No vertical exchange of momentum
T000ug08 TS02HD60F01_**src52** T012ug08 TS02HD60F01_**src52**	No vertical TS diffusion
T000ug08 TS02HD60F01_**src53** T012ug08 TS02HD60F01_**src53**	No horizontal TS diffusion
T000ug08 TS02HD60F01_**src54** T012ug08 TS02HD60F01_**src54**	No vertical TS advection
T000ug08 TS02HD60F01_**src55** T012ug08 TS02HD60F01_**src55**	No horizontal TS advection
T000ug08 TS02HD60F01_**src56** T012ug08 TS02HD60F01_**src56**	No vertical TS advection & diffusion
T000ug08 TS02HD60F01_**src57** T012ug08 TS02HD60F01_**src57**	No horizontal TS advection & diffusion
T000ug08 TS02HD60F01_**src58** T012ug08 TS02HD60F01_**src58**	No horizontal exchange of momentum, no Smagorinsky diffusion
T000ug08 TS02HD60F01_**src60** T012ug08 TS02HD60F01_**src60**	No vertical exchange
T000ug08 TS02HD60F01_**src61** T012ug08 TS02HD60F01_**src61**	No horizontal exchange

T000 = no OWF, T012 = 12-turbine OWF, ug08 = 8 m/s of prescribed geostrophic wind, TS01 & TS02 prescribed temperature and salinity field, HD60 = ocean depth of 60 m, F01 = only wind and atmospheric surface pressure forcing

5.2.1 Analysis of Dynamical Pattern Under Barotropic Conditions

Investigation of the ocean's reaction on operating OWF under barotropic conditions helps to strike a statement of the triggering of the vertical motion.

Generally, the barotropic conditions of an ocean system are mostly identified in the relative homogeneous deep layer. Physically, a barotropic ocean means parallelism of isopycnic and isobar surfaces having a constant slope with depth. The horizontal pressure gradient, as well as the geostrophic flow, is constant with the depth.

The 'simplification' of the baroclinic model to barotropic model is done by absence of baroclinic components. In the case of here used HAMSOM simulation *T012ug08 TS01HD60F01_BT (BT)*, barotropic means negligence of the baroclinic pressure component and the nonprognostic calculation of temperature and salinity. A better treatment of temperature and salinity and further eliminations of baroclinic components are not possible due to model design.

In this context, it must be mentioned that HAMSOM uses a semi-implicit numerical scheme. The pressure component only is separated into internal (baroclinic) and external (barotropic) components. Referred to Backhaus (1985), the separation of barotropic and baroclinic pressure components is indicated in following relation:

$$P(z) = (g\varrho_1\zeta)_{\text{ext}} + \left(P'(\zeta) + g\int_z^0 \varrho'\,\text{d}z\right)_{\text{int}} \tag{5.1}$$

with

$P(\zeta) :=$ atmospheric pressure at sea level; $P'(\zeta) :=$ atmospheric pressure anomaly at sea level;
$\rho_1 :=$ actual density of layer;
$\rho' := \rho_1 - \rho_0$, ρ_0 as reference density;
$g :=$ acceleration due to gravity;
$\zeta :=$ surface elevation;
ext := external component (barotropic);
int := internal component (baroclinic).

Based on explanations by Backhaus (1985), the atmospheric pressure $P(\zeta)$ is put into the internal pressure component because it does not need to enter the implicit scheme for the external pressure variations, which involves the sea surface elevation at the first layer. In the case of the internal component, the atmospheric pressure enters as a pressure anomaly $P(\zeta')$ due to the approximation of the internal pressure gradients. That approximation obtains a high accuracy when it depends entirely upon anomalies (Backhaus 1985).

The variations in the temperature and salinity field and thus in the density field occur at much lower frequencies than the oscillation of the free surface. Hence, they are solved by means of an explicit scheme, and therefore HAMSOM can only simulate temporal and spatial changes of the large baroclinic fields. But the use of implicit and explicit system components ends in the fact that a barotropic mode during the implicit scheme strongly influences the temperature and the salinity being treated in the explicit scheme. This means that the advection velocities derived from the solution of the primitive equations are centered in time between the adjacent time levels for heat and salinity because they are defined half a time step apart from these. Finally, a constant temperature and salinity field gives the barotropic conditions for the hydrography.

The pressure is incurred in the equation of motion. Vertically integrated over a depth range h, according to a computational model layer thickness h, the equation of motion can be formulated like in Backhaus (1985) as

$$\frac{\partial}{\partial t}\begin{pmatrix} U \\ V \end{pmatrix} + \begin{pmatrix} 0 & -f \\ f & 0 \end{pmatrix}\begin{pmatrix} U \\ V \end{pmatrix} + \frac{h}{\varrho}\begin{pmatrix} \partial p/\partial x \\ \partial p/\partial y \end{pmatrix} = \begin{pmatrix} X \\ Y \end{pmatrix} + \begin{pmatrix} \Delta\tau^x \\ \Delta\tau^y \end{pmatrix} \qquad (5.2)$$

with

$f :=$ Coriolis parameter;

$\begin{pmatrix} X \\ Y \end{pmatrix} :=$ for example advective terms, horizontal diffusion term;

$\Delta :=$ vertical difference;

$\tau :=$ shear stress term;

$U, V :=$ components of transport averaged over depth h;

$p :=$ pressure.

Finally, the barotropic HAMSOM run BP is adjusted by neglecting the baroclinic pressure gradient within the equation of motion (Eq. 5.2).

Based on this adjustment, HAMSOM run BP provides the following *results in case of barotropic conditions*:

In Fig. 5.9, the difference between OWFr and REFr of barotropic simulations BTM is depicted for ocean variable surface elevation ζ, vertical velocity component w in 3.0-m depth, and SST field after 24 h of operating wind turbines.

In case of barotropic mode, the change of surface elevation ζ and the vertical velocity component shows similar structures like the master run (baroclinic mode). As expected, the temperature field does not show an effect due to an OWF in the case of barotropic conditions.

The maximal difference in *surface elevation* ζ is $+3.65 \times 10^{-3}$ m; the minimal difference counts -6.16×10^{-3} m, which is slightly lesser than the master run,

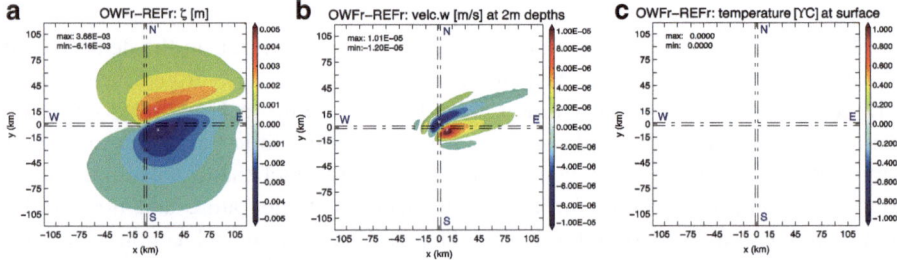

Fig. 5.9 12-turbine OWF effect (OWFr–REFr) on the ocean under barotropic conditions. Ocean variables are (**a**) surface elevation ζ, (**b**) vertical velocity component w at 2-m depths, and (**c**) SST. The existence of vertical motion in the model area and no reaction in the temperature field, respectively in the salinity and density field, due to OWF under barotropic conditions, leads to the assumption that vertical motion is a result of changed barotropic pressure implicated by surface disturbance of ζ due to operating wind turbines

which shows $+5.86 \times 10^{-3}$ and -9.16×10^{-3} m. Here, the comparison of the extreme values of ζ results in an about 35 % weaker ζ effect in the case of BTM, but in relation to the mean ζ effect, the BTM simulation is 4.54 % weaker than the master simulation.

The weaker effect on surface elevation can be explained using the *vertical velocity component*. The existence of the OWF effect on the vertical velocity component indicates that the vertical motion occurs as a cause of ζ dipole formation. The wind wake causes a wake in ocean flow, which provokes congestion of mass, and the downwelling is a compensating reaction.

The impact on vertical velocity component w by BTM has a maximum of 1.0×10^{-5} m/s and a minimum speed of -1.2×10^{-5} m/s in 3.0-m depth, Fig. 5.10. While here the minimum, downwelling, is nearly equal to the master

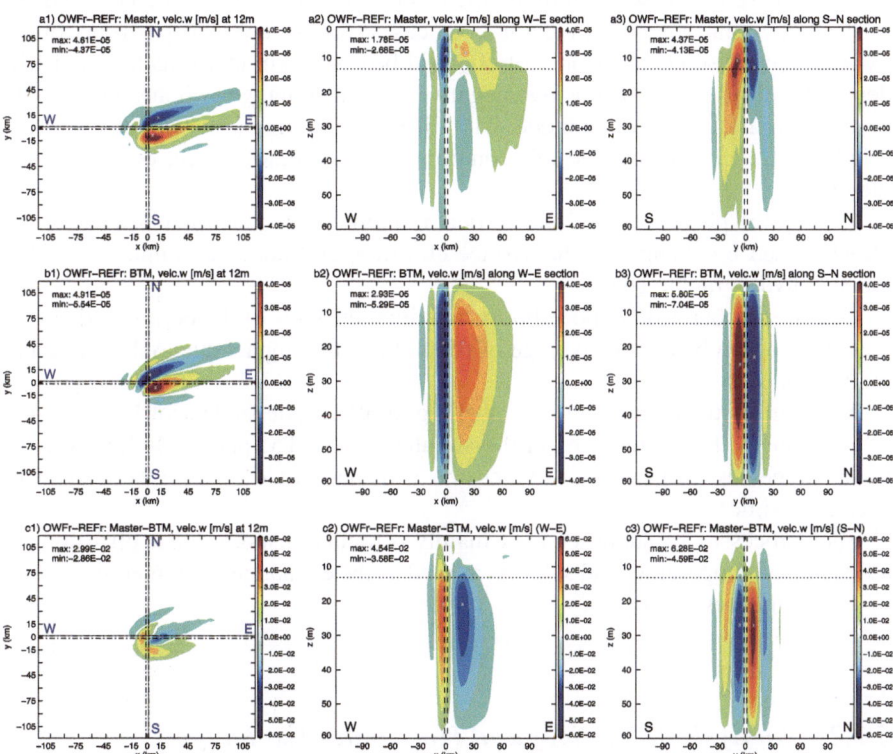

Fig. 5.10 12-turbine OWF effect (OWFr–REFr) on the vertical velocity component w for (**a1**)–(**a3**) master run, means barotropic mode (Master), for (**b1**)–(**b3**) run under barotropic conditions (BTM) and (**c1**)–(**c3**) the difference between Master and BTM. Results are presented in the horizontal at 12-m depths (**a1–c1**) and along the cross-sections W–E (**a2–c2**) and S–N (**a3–c3**) through the OWF. The OWF is placed around $x = y = 0$. BTM results in a smoother and intensified vertical velocity, especially in deeper depths and even in 12-m depths. In the case of Master, maximal changes are concentrated around the 12-m layer

run, having -1.1×10^{-5} m/s; the maximum, upwelling, of the barotropic run is 34.29 % lower than the master run.

The effect of the barotropic run, compared to the baroclinic master run, is, under barotropic conditions, stronger at the thermocline by around 3.0×10^{-5} m/s, which is equivalent to a 45 % stronger effect for upwelling and downwelling by BTM, Fig. 5.10. Stronger vertical mixing over time in the barotropic case results in a slightly faster compensation of the surface elevation. Thus run BTM shows lower values after 1 day at surface.

The vertical velocity component of the barotropic run has its maxima in 27.00-m depth with values of 5.0×10^{-5} up to 7.0×10^{-5} m/s (~6 m/d). Here, the vertical cells around the OWF are more intensified with the depth than in the master run, but their horizontal dimensions are restricted to 15 km, compared to the 30 km in the master run along the S–N cross section through the OWF, Fig. 5.10. The cells in BTM are smoother and more symmetric than the ones in the master run, especially along the W–E cross section through the OWF. Therefore, the positions of extrema in the horizontal are not equal for both cases. At surface, the positions of the extrema have discrepancies only of one grid box, so 3 km in x-direction, but with depth, the positions of the positive/negative maximal change switches more to the north/south in the barotropic mode, with difference to the master run of 3 km (one grid box). Hence, under barotropic conditions, the extreme changes occur closer to the OWF.

Finally, the physics behind the vertical motion can be identified as a barotropic effect caused by changes in the surface elevation and not as an impulse of OWF induced hydrographic changes.

Overall, the barotropic mode boosts maximal changes of the ocean system over the whole ocean box by an average of 65 %; thus, the impact due to the baroclinic mode counts 35 %.

The start of upwelling and downwelling at all is independent of hydrographic OWF changes, but differences in the simulation of BTM and BCM link to additional processes triggering the dimension and magnitude of the OWF effect on ocean dynamics.

Such processes are supposed to be mainly vertical transports due to diffusion and advection of temperature and salinity, as well as the exchange of momentum, which in turn influences diffusivities.

5.2.2   Analysis of Vertical and Horizontal Exchanges

Reflection of the OWF effect on the ocean under barotropic conditions yields to the result of vertical motion being a requirement of mass redisturbance based on wind wake and reduced surface flow. But besides the barotropic factor, additional processes have to be considered to describe the final phenomenon on the ocean impacted by the OWF.

Such processes are defined for horizontal and vertical exchanges comprising the exchange of momentum, heat, mass, and salinity. They can be summarized under advection and diffusion. The exchange processes are based on local gradients, which are strengthened by turbulent motion.

In hydrology, advection means the transport of solved or suspended material in water with the flow, thus with the mean speed and direction of the ocean flow (Rubin and Atkinson 2001). In the vertical, salinity and heat transport, as well as the one of momentum, is related to the vertical velocity component w.

Diffusion is defined as a transport of molecules along a concentration gradient (Jones 2010); as intensified, the gradient is as stronger as diffusion.

The different terms regarding advection and diffusion in HAMSOM were switched on and off by manipulating the source code to evaluate their impact factor. Therefore, some explanations about HAMSOM details are necessary. The following description is in accordance with Backhaus (1985) and Pohlmann (2006).

HAMSOM's diffusion and advection terms are bounded in the equation of motion, the transport equation of temperature, respectively salinity. While the diffusion terms, the vertical shear stress and the terms determining the surface gravity waves are formulated implicitly, all other terms are formulated explicitly (Pohlmann 2006). The advective terms in the momentum equation and the transport equation for temperature and salinity are solved explicitly, with the exception of the vertical advective term. Advection and diffusion of temperature and salinity in HAMSOM follow a method related to the difference scheme of second order accuracy from Lax and Wendroff (1960). That second order advective scheme was implemented in HAMSOM in combination with a superbee flux limiter (Roe 1986) by Hein (2013) and is able to simulate the diffusion more realistically, and the mixing processes are fully controlled by physical processes. The use of the flux limiter avoids spurious oscillation due to shock waves, contact with surfaces, or discontinuous derivatives across any characteristics (Roe 1986). Fick's laws describe the diffusion itself.

At this point, two important coefficients of the model HAMSOM must be introduced—the vertical eddy viscosity coefficient A_{vc} and the vertical eddy diffusion coefficient A_{dc}.

The coefficient A_{vc} is used because small-scale vortices or eddies in the motion cannot be resolved at the mesoscale. To consider such vortices, the large-scale motion is calculated with eddy viscosity that characterizes the transport and dissipation of energy in the smaller scale flow. So A_{vc} is linked with the transfer of momentum caused by turbulent eddies and is linked to the molecular exchange in the vertical because the vertical eddy diffusion coefficient A_{dc} in the prognostic equation of temperature is calculated via A_{vc}. In an equation, this subject matter is described as follows.

A_{vc} can be separated into different applications, into the usage in surface mixed layer, the bottom mixed layer, and the interior, which is under laminar conditions. In the case of the laminar part, the coefficient A_{vc} is defined as 0.0134 cm^2/s. In the case of the turbulent part, the coefficient A_{vc} is defined by

$$A_{vc} = (c_{ML} \times h_{ML})^2 \times \sqrt{\left(\frac{\partial u}{\partial z}\right)^2 + \left(\frac{\partial v}{\partial z}\right)^2 + \frac{1}{S_M \rho} \frac{g}{\partial z} \frac{\partial \rho}{\partial z}} \qquad (5.3)$$

with

$c_{ML} := 0.05$,
$h_{ML} :=$ thickness of mixed layer,
$S_M :=$ turbulent Schmidt–Prandtl number ~1.3.

The vertical eddy diffusion coefficient A_{dc} is described through the generally accepted linear relation (Pohlmann 2006):

$$A_{dc} = \frac{A_{vc}}{S_M} \qquad (5.4)$$

Both parameters, the vertical eddy viscosity and diffusion coefficient, are necessary to parameterize the Reynolds stress terms in the shallow water equations and in the transport equation for heat (Pohlmann 2006). So A_{vc} impacts the vertical exchange. Due to that, it was possibly necessary to examine the influence of this coefficient on the influenced ocean by keeping A_{vc} minimal. Minimal means the usage of a laminar A_{vc}. That procedure helps to define its impact on the vertical velocity component and the distribution of temperature and salinity through the vertical diffusion.

In the following, the analysis of the vertical exchange is presented, followed by the analysis of horizontal exchanges.

Vertical Exchange
This sensitivity study aims to estimate the impact of vertical and horizontal diffusion and advection on the new state of system after 1 day.

The analysis of the vertical exchange comprises, next to the normal run used as a reference (*src50*), five simulations, labeled as *src60, src51, src52, src54*, and *src56*, including different adjustments regarding vertical exchange. These adjustments consider the aforementioned parameters vertical eddy viscosity coefficient A_{vc}, vertical eddy diffusion coefficient A_{dc}, and vertical advection term of temperature and salinity (labeled here as TSVA) in their prognostic equations.

Three options regarding vertical exchange were applied:

- *Setting A_{vc} to a minimum* of 0.0134 cm^2/s nearly avoids vertical exchange of momentum. Subsequent, the momentum at surface increases, so the horizontal velocity increases because lower layers will not be forced by momentum of surface layer. The surface motion due to shear stress triggers the lower layers.
- *Setting A_{dc} to zero* avoids vertical diffusion of temperature and salinity (TS). A_{dc} is zero in case of a minimal A_{vc} due to its definition. That means no vertical mass transport of TS.
- *Setting TSVA to zero* avoids vertical transport of temperature and salinity with the vertical flow, thus with *w*-component.

In the following, these three options were applied to HAMSOM in different combinations. In the first sensitivity run, *src60*, all three options are set to minimum/zero. In a second step, in the sensitivity run *src51*, only A_{vc} is set to minimum; vertical advection and diffusion are treated as normal. In the sensitivity run *src52*, consequences of vertical diffusion are carved out by setting A_{dc} to zero; in *src54*, TSVA is neglected, and in *src56*, TSVA and vertical TS diffusion are set to zero.

For analysis, extrema of variables along the N–S cross section after 1 day of simulation are used as representative data set. The maximal effect on the ocean due to an operating OWF is illustrated, separated into positive effect, means an increase of variable compared to reference run without wind turbines, and opposite effect— the negative effect. Based on the asymmetric surface elevation, the maxima and minima are not symmetric. An overview of changes in the OWF effect on the ocean by vertical exchange processes is pictured in Fig. 5.11.

The sensitivity run, src60, preventing vertical exchange of momentum, vertical advection, and diffusion of temperature and salinity (TS), shows a weaker effect in surface elevation due to a stronger impact on velocities but does not show an effect in the hydrographic stratification, and the thermocline does not form an excursion around the OWF. Higher velocities are based on a constant hydrographic field unpersuaded by additional changes in the density and so the pressure field.

Figure 5.12 illustrates the difference in temperature along the cross section from S to N through the OWF for the normal simulation src50 and the run src60 without vertical exchange processes. Runs without vertical TS exchange processes have a warmer upper layer because the exchange of heat is forbidden, while in the normal run, the upper layers become cooler due to the exchange via the thermocline. The other way around is also explained by the exchange via the thermocline, which ends in warmer layers close below the thermocline.

Ignoring vertical diffusion, src52, the vertical exchange of temperature and salinity (TS) is connected with vertical advection. Results show that the negligence of the diffusion increases the effect on hydrographic conditions, compared to the normal run (Fig. 5.11). On one hand, that leads to the assumption that vertical advection plays an important role for the exclusion of the thermocline, and on the other hand, the vertical diffusion seems to break the development of the hydrographic effect. Having a look at the distribution of the temperature in comparison with the normal run in Fig. 5.13, less obvious changes are identifiable. The examination of the OWF effect illustrates that a cooling occurs at and above the thermocline southerly of the OWF and the warming occurs only below the thermocline northerly of OWF. The change is sharp, in the form of small 'arrows,' linked to the direction of the vertical velocity component w, it is stronger limited in the vertical than in the normal run. So the diffusion supports reduction of the gradients over the vertical layers. Therefore, in the normal run, the effect has a more oval form blurred over more layers. Hence, differences between run src52 (no vertical diffusion) and the normal run are located at thermocline ±30 km around the OWF along the S–N section. Here, we can say that the diffusion does not cause the exclusion of the thermocline but triggers the form and therewith the extrema of the OWF effect on temperature.

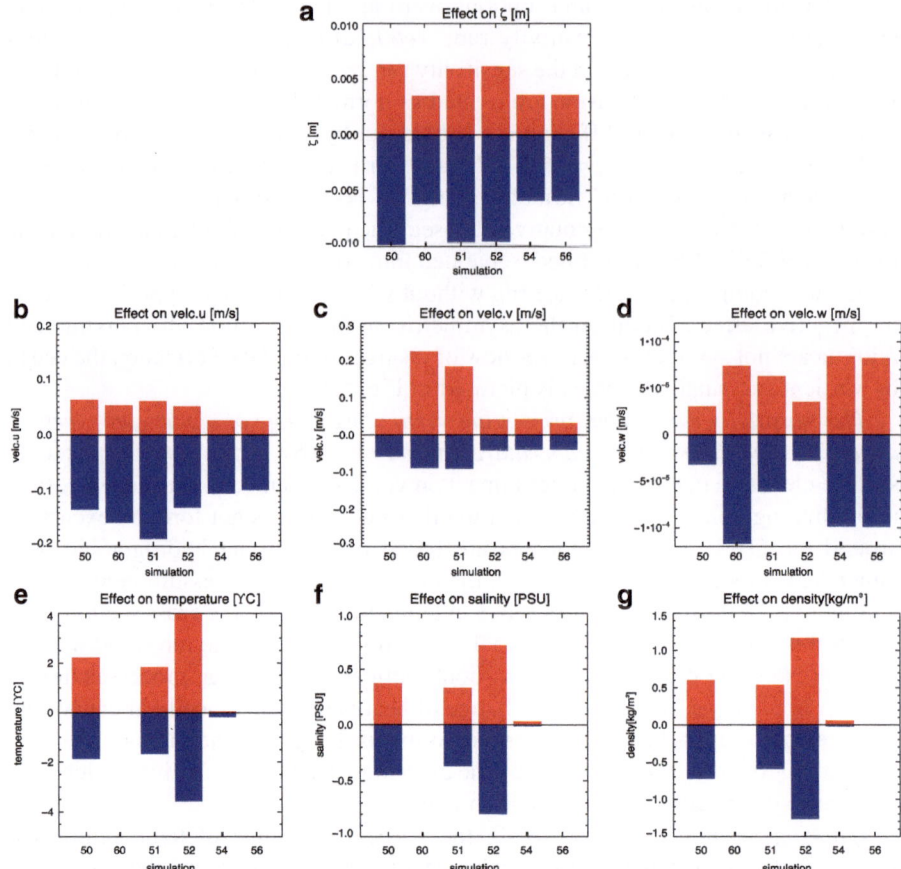

Fig. 5.11 Overview of maximal OWF effect (OWFr–REFr) on the ocean after 1 day of simulation along cross-section S–N in comparison with sensitivity runs regarding the vertical exchange. *Y*-axis gives the change of the variables (**a**) surface elevation, (**b**) velocity component *u*, (**c**) velocity component *v*, (**d**) velocity component *w*, (**e**) temperature, (**f**) salinity, and (**g**) density. *X*-axis comprises simulations of sensitivity study, src*: 50 denotes 'normal run'; 60 is run without vertical exchange of momentum, advection, and diffusion; 51 ignores vertical exchange of momentum in simulation, 52 ignores vertical diffusion, and 54, vertical advection; 56 denotes a run ignoring vertical diffusion, as well as vertical advection, but vertical exchange of momentum is treated as normal

The vertical dimension is halved, and the effect magnitude of the hydrographical variables is increased by an average of 45.82 %, compared to the normal run.

In a third step, the *vertical TS advection is set to zero, src54*, which prevents the exchange of heat with the vertical motion *w*. The effect on the hydrographic values is minimal but is exalted in the upwelling regions, which means a decrease of temperature and increase of salinity (Fig. 5.14).

Compared to the normal run, the effect is around 95.0 % smaller for the hydrographic variables, which means nearly no change in TS is registered. So the

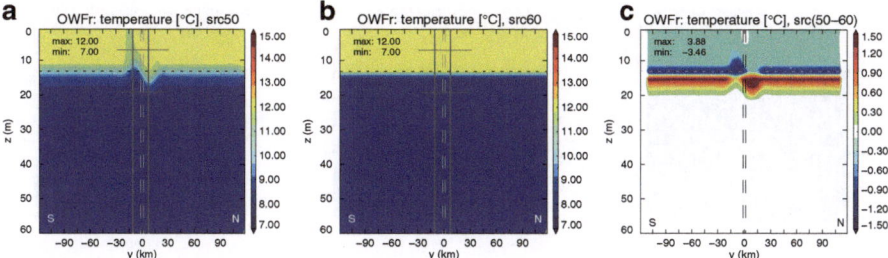

Fig. 5.12 Comparison of the OWFr temperature stratification along the S–N cross section through 12-turbine OWF around $P(0,0)$ between (**a**) the normal run (src50) and (**b**) the sensitivity run, avoiding vertical exchange of momentum, vertical diffusion, and advection (src60) after 1 day of OWF operation. The vertical changes in the hydrographic conditions are forbidden in src60. Illustration (**c**) shows the difference between src50 and src60. The *dashed horizontal line* marks the depth of the thermocline, the *dashed–dotted lines* mark the OWF area, and the *solid lines* accent the effect dimension

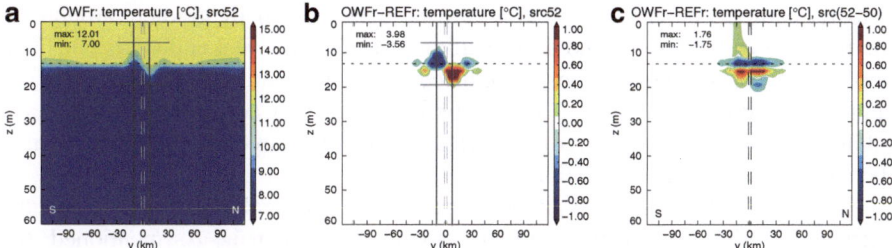

Fig. 5.13 OWF effect on the ocean's temperature stratification based on sensitivity run without diffusion (src52) for (**a**) OWFr and (**b**) difference between OWFr and REFr and (**c**) comparison of the effect with the normal run (src50) after 1 day of OWF operation. No vertical diffusion means that the occurred changes refers to the vertical advection. The *dashed horizontal line* marks the depth of the thermocline, the *dashed–dotted lines* mark the OWF area, and the *solid lines* accent the effect dimension

vertical advection plays a dominant role for the development of the thermocline exclusion.

A comparison of src54 (no vertical advection) with the *simulation, without vertical advection, and without vertical diffusion* (src56) shows that the diffusion ends in a diffusive thermocline with a linear transition between the upper and lower water layers (Fig. 5.14). Without vertical advection and diffusion, a sharp transition exists. Hereby, it becomes apparent that the diffusion supports changes within the OWF in upwelling direction via the thermocline. Therefore, the diffusion also supports a temperature increase from surface down to the thermocline following the gradient of concentration; as known, diffusion acts again as concentration gradient.

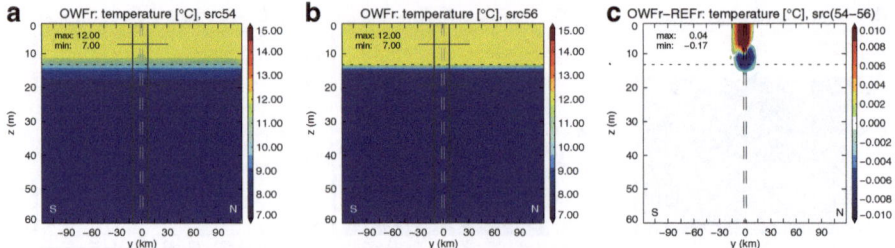

Fig. 5.14 Impact of the vertical diffusion (run src54: neglecting vertical advection) on the (**a**) OWFr temperature stratification, compared to (**b**) the sensitivity run without advection and diffusion (src56 after 1 day of OWF operation; the difference is pictured in (**c**). The diffusion leads to a diffusive transition at the thermocline but without a sharper thermocline. The heat transport is supported by diffusion within the OWF. Sensitivity run src54, in comparison with src56, depicts the single effect of vertical diffusion. The *dashed horizontal line* marks the depth of the thermocline, the *dashed–dotted lines* mark the OWF area, and the *solid lines* accent the effect dimension

For this reason, the vertical advection dominates the TS effect based on the velocity component w. But with time, the temperature/density gradient around the thermocline within the up- and downwelling cells become weaker due to intensified vertical advection, which leads to reduced vertical velocities in the normal run (Fig. 5.11). Without vertical TS advection, the up- and downwelling cells have intensified magnitudes of extrema, which are in average three times stronger than in the case of the normal simulation.

Figure 5.15 illustrates the single impact of the vertical TS diffusion and the vertical TS advection on the vertical velocity component w. As mentioned, the vertical advection reduces up- and downwelling by an average of 64.38 %, while diffusion only supports the vertical motion by around 1.21 % in relation to src56 (no vertical TS advection and diffusion).

Summarizing, the OWF effect on the hydrography based on vertical advection and diffusion acts by the same means but a difference in the velocity field, mostly contradictory.

The last vertical exchange mode here, simulation run *src51*, considers HAMSOM vertical eddy viscosity coefficient A_{vc} and so the *vertical exchange of momentum*.

Regarding hydrography, coefficient A_{vc} increases the OWF effect by 10.94 % for the negative effect and by 17.51 % for the positive effect, which means an average impact of 14.23 % (Fig. 5.11). The vertical velocity component w is greater in the sensitivity run than in the normal run, exactly by 49.45 %, as well as the v-component with a 77.03 % greater increase and the u-component with a wake increase of 28.25 % (Fig. 5.11). Generally, A_{vc} supports changes in hydrographic fields due to stronger vertical motion and triggers the dimension of the wake in the velocity field. These results here also strengthen the thesis of the vertical motion having its origin in changed barotropic conditions. Minimizing the vertical eddy viscosity, coefficient A_{vc}, almost neglects the vertical exchange of momentum,

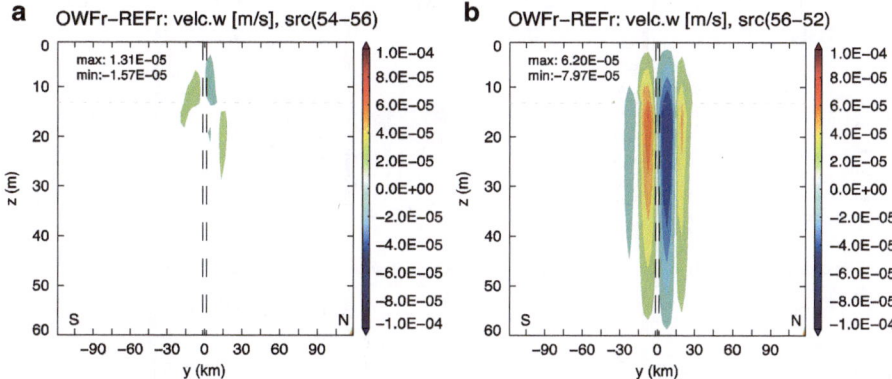

Fig. 5.15 Impact of diffusion (**a**) on velocity component w and impact of advection (**b**) on velocity component w after 1 day of OWF operation by taking the differences between src54 (no vertical TS advection), src52 (no vertical TS diffusion), and src56 (no vertical TS advection/diffusion). The diffusion supports up- and downwelling a little bit, while advection leads to control vertical motion and reduces magnitudes. The OWF effect on w is stronger in the case of no advection; in the case of diffusion, the effect is similar, compared to the normal run

which dominantly influences the upper layers. Based on the definition of the vertical eddy diffusion coefficient, depending on A_{vc}, the vertical diffusion of TS is avoided in the case of a minimal A_{vc}. But due to a less impact of the diffusion on the final OWF effect on the ocean, that side effect will not much influence the manner of A_{vc} impact.

The sensitivity run src51 (no vertical exchange of momentum) leads to an increased reduction of the velocity component u, while the velocity component v is strengthened, compared to the normal run. The strong change in the horizontal velocity is explained on one hand by a reduced Ekman transport due to the wake and by a neglected transmission of momentum from the surface layer to the layers below. Therefore, the velocity wake area becomes more intensified, as well as the wake flank area. A more intense wake leads to a stronger effect in surface elevation, which triggers a change in the v-component because of a reduction in the Ekman transport (see explanations to Ekman transport in Sect. 5.2.3).

Figure 5.16 pictures the horizontal velocity components, the direction of the horizontal velocity field at the surface, and the vertical velocity component w along the cross section from S to N through the OWF for sensitivity run src60 (no vertical TS diffusion & advection and no vertical exchange of momentum), sensitivity run src56 (no vertical TS diffusion & advection), and the difference between the two to capture the single impact of a normal-handled A_{vc} coefficient after 1 day.

The direction of flow differs due to the difference in the velocity components and hence in the Ekman transport. The horizontal velocity field in a run without vertical exchanges (src60) causes a more intense vertical velocity component due to a more dominant gradient in the horizontal velocity field.

Using a *minimal A_{vc} and also ignoring the vertical TS diffusion and advection (src60)*, the vertical component w ends in the strongest downwelling of all

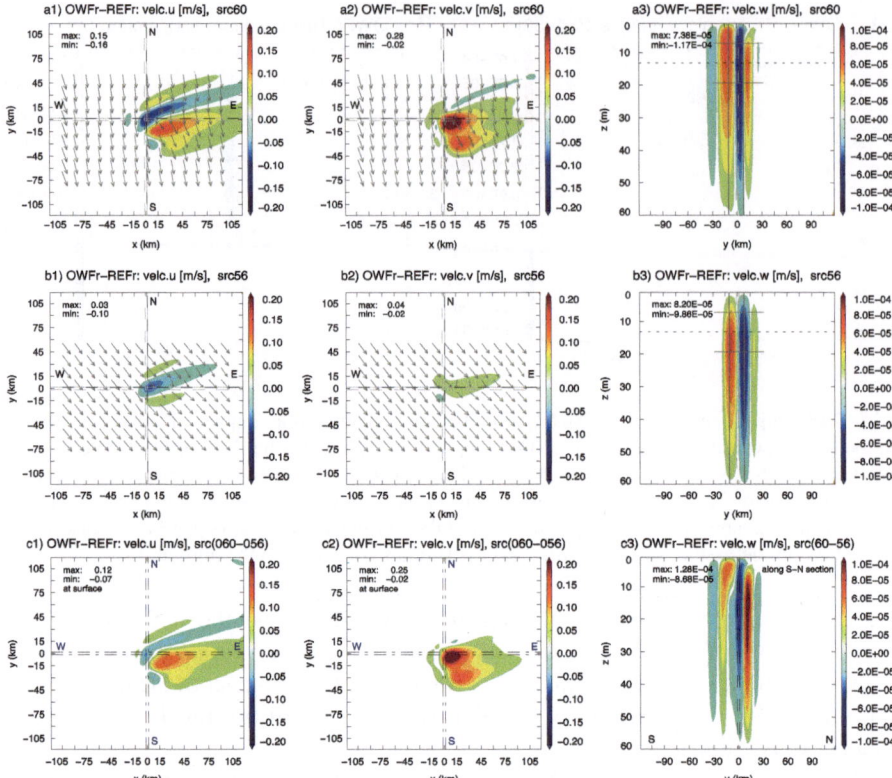

Fig. 5.16 Single impact of A_{vc} on velocity components independent of vertical diffusion and advection, after 1 day of OWF operation, at 2-m depths. Illustrations (**a1**)–(**a3**) show results for run src60 (no vertical exchange, A_{vc} = min, no TS advection/diffusion), (**b1**)–(**b3**) show results for run src50.56 (no vertical TS advection/diffusion), and (**c1**)–(**c3**) show the difference between src60 and src56. Variables are (**a1**)–(**c1**) velocity component u, (**a2**)–(**c2**) velocity component v in the horizontal, and (**a3**)–(**d3**) velocity component w over the vertical along cross-section S–N through the OWF. *Arrows* define direction of horizontal velocity field. A_{vc} controls horizontal dimension of the OWF effect. Therefore, a clearly defined wake with flanks is adopted from the wind field, and therefore the vertical motion is limited to two main cells. Additional vertical cells are suppressed due to a weaker gradient at the surface

sensitivity runs (Fig. 5.16), whereas the vertical motion cannot have an effect on temperature due to negligence of vertical TS advection/diffusion.

Differences with the normal run count 72.82 % for downwelling and an increase for upwelling of 58.53 %.

Considering the run *without diffusion and especially advection, src56,* (stronger advection triggers w), the magnitudes of up- and downwelling cells are more symmetric by maximum average changes to the normal run of a 49.44 % increase.

If we prohibit a vertical exchange of the momentum, then the vertical motion, which is triggered by the surface elevation and the horizontal velocity, can affect lower layers more easily because the effect at upper layers run faster due to a stronger gradient in momentum. The stronger gradient and an intense horizontal velocity depend on the fact that the momentum cannot be transferred from the top layer to the layers below. While advection triggers the magnitude of w, A_{vc} also controls the number of vertical cells. Additionally, the vertical cells, besides the two main cells around the OWF, are caused by velocity gradients at the surface and are suppressed in the case of normal vertical eddy viscosity coefficient A_{vc}.

Summarizing, the vertical exchange triggers dimension and magnitude of the up- and downwelling cells. The hydrographic conditions are influenced by the vertical advection, which again affects the vertical motion. The vertical diffusion acts especially at the thermocline and support an exchange within the OWF district. The vertical eddy viscosity coefficient A_{vc} affects the vertical exchange of momentum and the vertical velocity component w due to variations in the form and the magnitude of the wake in the velocity field. The flanks of the wake become more important, and the v-component increases in direction from north to south. Also, the wake is intensified, as well as the vertical velocity component w. As higher A_{vc} is stronger, the velocity components are reduced at the surface.

Horizontal Exchange

The horizontal exchange plays a secondary role for the OWF effect on the ocean system in the vertical. The sensitivity runs regarding horizontal exchanges are listed in Table 5.1.

Figure 5.17 illustrates the extrema along the cross-section S-N through the OWF for each sensitivity run regarding horizontal exchanges like diffusion, advection, momentum, and a combination of any of them. Overall differences between the various horizontal exchange modes exist, but the extrema do not strongly vary with maximal discrepancies of ± 10 %, compared to the normal run (src50), with the exception of src53 and additional src58 for the velocity components. Sensitivity run src53 avoids horizontal TS diffusion, while src58 avoids Smagorinsky diffusion and exchange of momentum.

The Smagorinsky diffusion describes a nonlinear diffusion acting horizontally, depending on u- and v-components. The Smagorinsky diffusion coefficient K_{smg} includes the horizontal tension strain T_{hs} and the horizontal shearing strain S_{hs}.

In Cartesian coordinates, it can be written as follows:

$$\frac{\partial u}{\partial t} + \vec{v}\,\Delta u = \ldots + K_{smag}\Delta u \tag{5.5}$$

$$\frac{\partial v}{\partial t} + \vec{v}\,\Delta v = \ldots + K_{smag}\Delta v \tag{5.6}$$

$$\frac{\partial w}{\partial t} + \vec{v}\,\Delta w = \ldots + 0 \tag{5.7}$$

with $K_{smag} = l_s^2 \sqrt{T_{hs}^2 + S_{hs}^2}$ and $T = \frac{\partial u}{\partial x} - \frac{\partial v}{\partial y}$, $S = \frac{\partial u}{\partial y} + \frac{\partial v}{\partial x}$.

The use of the Smagorinsky horizontal diffusion stabilizes the dynamical core against horizontal shear instabilities.

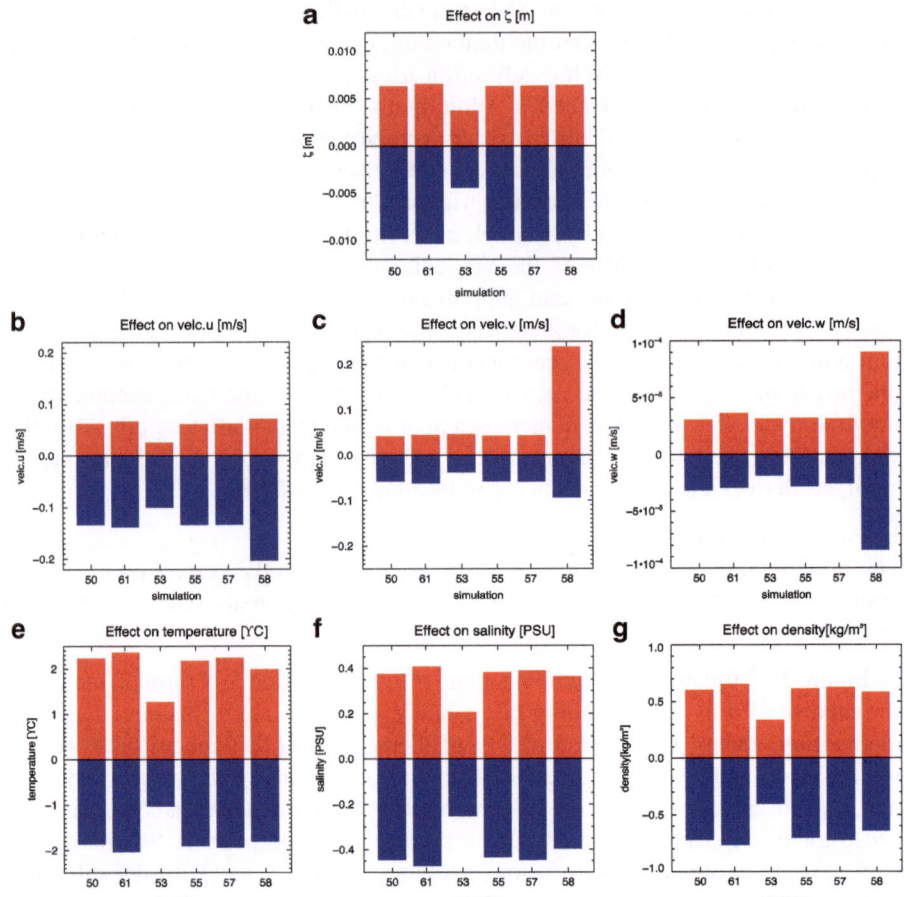

Fig. 5.17 Overview of maximal OWF effect on the ocean after 1 day of OWF operation along cross-section S–N in comparison with sensitivity runs regarding the horizontal exchange. Y-axis gives the change of the variables (**a**) surface elevation, (**b**) velocity component u, (**c**) velocity component v, (**d**) velocity component w, (**e**) temperature, (**f**) salinity, and (**g**) density. X-axis comprises simulations of sensitivity study src.*: 50 denotes 'normal run'; 61 is run without horizontal exchange of momentum, advection, and diffusion; 53 ignores horizontal diffusion; 55 ignores horizontal advection; 57 avoids horizontal advection and diffusion; and 58 is no use of Smagorinsky (HAMSOM parameter horcon = 0)

Therefore, the negligence of Smagorinsky diffusion leads to a stronger wake and hence stronger changes in the velocity components in src58, compared to normal run (Fig. 5.17). Changes in the horizontal velocity field again affect the vertical eddy viscosity coefficient A_{vc}, which impacts the vertical velocity component and the hydrographic variables.

The reduction of the hydrographic conditions in the case of no horizontal TS diffusion (src53) shows that the horizontal diffusion intensifies the gradients in the density field and thereby impacts velocity field and surface elevation.

Here (in src53), the effect on the surface elevation develops very slowly, compared to the normal run, and so the gradient is weaker from the beginning on, which again weakens all changes in the ocean. Without horizontal diffusion, the rise in surface elevation due to the velocity wake is more locally limited and not spread over the whole area and is controlled by the horizontal advection.

Summarizing, horizontal exchanges balance the OWF effect and the vertical structure of the ocean system. Horizontal exchanges control gradients in the horizontal, which affects vertical changes. Especially, horizontal diffusion (TS diffusion and Smagorinsky diffusion) influences the final magnitude of OWF effect, but dominantly, the vertical processes trigger the OWF effect on the ocean system.

5.2.3   Assessment and Integration of Effect Analysis

Previous documented analyses deal with the OWF effect on the ocean system under barotropic and baroclinic conditions and in the case of various exchange process combinations. The manner of the vertical motion is related to changes in the barotropic pressure due to the change in the surface elevation released by the flow reduction due to the wind wake. The treatment of the exchange analysis results in the statement that especially vertical advection with vertical motion triggers the change in the hydrographic conditions. Partly betoken during prior explanations, this section illustrates the main physical principle behind the changes that occurred on the ocean system.

Starting once more from an initial situation, our ocean system is forced by a constant wind field, which is affected by a wind farm. The wind turbines of the wind farm detract the atmosphere's energy by transforming wind energy into a mechanical one. That energy detraction means a reduction of wind speed downstream of wind farm. So a wind wake is formed by OWFs. That wind field acts with the ocean surface and creates a surface stress.

Under undisturbed conditions, the constant wind field causes an Ekman transport. The Ekman transport is known as the net motion of water as a result of the balance between the Coriolis and turbulent drag forces. The net sum of the water column is theoretically ~90° directed to the movement of the wind (in the northern hemisphere), at least partly for real conditions.

Under operating OWF conditions, the wind wake causes a wake in ocean flow, and the locally reduced surface stress results in a reduced Ekman transport. This again causes a convergence of water masses within the wake and to the left of the wake and a divergence at the right side of the wake (looking into wind direction). The convergence/divergence of the water masses is associated with an increase/reduction of the sea level, which in turn induces downwelling/upwelling.

A commonly known connection between up- and downwelling and Ekman transport is in relation to coasts. An up- and downwelling also occur, besides coats, in the open ocean where winds cause the surface water to diverge from a

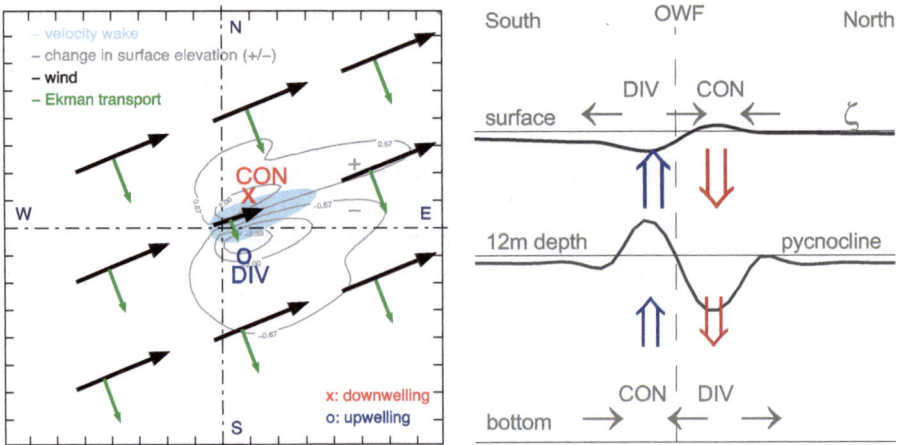

Fig. 5.18 *Left*: Schema of adjustment of Ekman transport resulting in up- and downwelling. Ekman transport (*green arrows*) is detracted to wind direction by theoretical ~90° to the wind (*black arrows*). Due to velocity wake (*light blue area*), Ekman transport is reduced, which causes convergence/divergence of water masses (change in surface elevation ζ is *light gray*) and downwelling 'x' and upwelling 'o.' *Right*: Schema of conditions of OWF-affected ocean system along cross-section S–N through the OWF. Constraint for downwelling is a positive decline in surface elevation ζ and convergence. The opposite is essential for upwelling. The pycnocline shows the opposite change of surface elevation

region or to converge toward some region. The last one is on hand here. Figure 5.18 schematically illustrates the term of conditions. Here, the horizontal velocity and Ekman transport are reduced within the area of the wake and increased surface elevation. The area of the velocity wake can be treated as a barricade or 'coast,' and now downwelling occurs where Ekman transport moves surface waters towards the region of velocity wake ('coast'); the water must pile up and sink. On the other side of the velocity wake, upwelling occurs where Ekman transport moves surface water away from the wake area (coast); surface water is then replaced by water that wells up from below. Upwelling and downwelling illustrate mass continuity in the ocean; that is, the change in distribution of water in the ocean area is accompanied by a compensating change in water distribution in another area. And those two areas are the dipole formation of the surface elevation northerly and southerly of OWF. The formation and dimension of surface elevation's dipole is a result of the Coriolis effect and the wind wake due to geostrophic conditions.

Finally, the vertical motion is a mandatory constraint of the wind-driven change in pressure (barotropic effect), in the velocity field, and so in the Ekman transport and can be defined as a wind-driven but coastal independent upwelling/ downwelling. The final dynamical situation of the model area around the OWF is illustrated in Fig. 5.18. A positive change in surface elevation means a lowered pycnocline, respectively the opposite for negative surface elevation. The zone where the upper part of the water column has a lower density is characterized by an increased sea surface height and a deepened pycnocline. At surface convergence is incurred that support downwelling, while divergence occurs at low surface elevation with upwelling. Due to divergences and convergences, the change in the horizontal velocity field at the surface varies from the wind field by time.

Based on exchange study, changes of hydrographic is a result of vertical motion, divergence, and convergence, dominantly supported by vertical advection.

External impact factors triggering the intensity of wind wake, and so of the ocean's answer, are considered in the analyses of the next section.

5.3 Analyzing OWF's Effect on the Ocean Under Various Assumptions

Analysis of the OWF's effect on the ocean under various assumptions aims to capture possible comprehensive impacts triggering wind wake and so the effect on the ocean.

On one hand, an exalted impact on the OWF's effect on the ocean considering aspects of influences due to duration of operating wind turbines, magnitude of wind speed, and size of OWF is documented.

On the other hand, different computational assumptions for ocean simulation are considered, focusing on additional aspects based on assumptions regarding forcing and design of ocean box.

In this connection, focus is more precisely the analysis

- 5.3.1: of the consistency of the ocean's effect due to duration of operation;
- 5.3.2: of wind speeds triggering wind wake magnitude;
- 5.3.3: of wind park power regarding the amount and arrangement of wind turbines;
- 5.3.4/5.3.5: of influence by forcing comprehending inducement by the Broström approach (5.3.4) and full possible meteorological forcing (5.3.5), which can be applied in HAMSOM;
- 5.3.6: of the design of ocean box dealing with a reduction of ocean depth from 60 to 30-m water depth.

The results are based on simulation under TOS-01 (model ocean box), considering the above-mentioned modifications using temperature and salinity start field TS01. Modifications include usage of forcing (F01, F02, F03, F04) and change of wind speeds (UG05, UG08, UG16), number of wind turbines (T12, T48, T80, T160), ocean depth (HD60, HD30), and horizontal resolution (HR3, HR1).

5.3.1 Analyzing Consistency of OWF Effect on Ocean

During the description of the common effect of a wind farm wake on the ocean, it was shown that the effect on the ocean appears immediately after using wind forcing with operating wind turbines. The theoretical analysis underlines that the first vertical cells come up triggering changes of hydrographic parameters. Therefore, what happens to the vertical velocity due to wind farm operation is analyzed. Offshore wind farms need special services, which can be only done in case of stagnancy of the rotor blades. Additionally, the wind turbines have an operating limit based on wind speed. Depending on the machine, a normal operating turbine is working between 5 m/s and up to 25 m/s (Vestas Wind Systems 2013), in the case of METRAS simulation between 2.5 and 17 m/s. The limited operation is caused by technical limitation. Therefore, we can assume that a wind farm or single wind turbine is not always in operating state. What will happen if we switch off turbines for a relatively short time duration, what happens in the case of weak or strong wind speeds, and what differences can we estimate between different wind park sizes?

Corresponding simulation for result presentation is *T012ug08 TS01HD60F03*.

The aim of this section is it to make a statement on how we can trigger ocean dynamics by switching on and off wind farms. Does the effect suddenly disappear, or is it slowly disappearing over longer time? How long will it take to bring ocean dynamics back to reference conditions?

Therefore, three OWF operation cases are used for this analysis. An overview of cases is given in Fig. 5.19, including different durations of turbine rotation. Thereby, 'on' means the usage of rotor disc approach in METRAS; 'off' means no use of the rotor disc approach, but as mentioned in Sect. 3.1.2, the wind field is affected because the frictional resistance of rotor disc is considered. Additional 'off_ref' is implemented that fully ignores the OWF existence.

Forcing is based on *M_T012ug08*onoff*, whose results are introduced in Sect. 4.3. The importance of forcing includes these three:

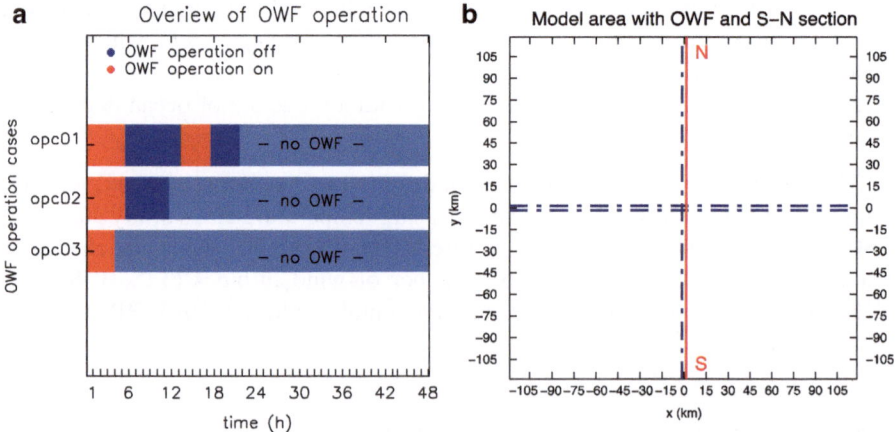

Fig. 5.19 Illustration (**a**) shows operation cases opc01, opc02, opc03 and their operating time. *Blue* is nonoperating wind turbines, *red* is operating wind turbines, and *cornflower blue* marks simulation where OWF is ignored completely. Illustration (**b**) shows *y*-section (S–N cross section) through the OWF for analysis (*red line*)

During 'on,' wake is developed and increased; at 'off,' wake is advected with the main wind field; and at 'off_ref,' the wind field is set to the reference run.

Operation case 01 (opc01) starts with 4.2 h of operating wind turbine, followed by 7.8 h 'off,' 4.2 h 'on,' 4.2 h 'off,' and finally 26 h 'off_ref.'

Operations case 02 (opc02) starts with 4.2 h 'on,' 6.1 h 'off,' and finally 36.5 h 'off_ref.'

Operation case 03 (opc03) starts with 2.6 h of operating wind turbines and is then switched off and uses 44.1 h 'off_ref.'

The analysis concentrates on the cross section from south to north through the OWF; see Fig. 5.19.

Figure 5.20/Figs. 5.21 and 5.22 summarize the development of the OWF effect based on the three operation cases opc01, opc02, and opc03 for velocity field and hydrographic conditions along cross-section S–N through the OWF.

Overall the Hovmöller diagrams of the three operation cases strongly differ from the previous long-time analysis (Sect. 5.1.2), which is based on a constant wind forcing per time step. Due to the turning on and off of the OWF, an additional side effect occurs–an inertial oscillation.

In all three cases, it was not possible to bring the ocean back to dynamical conditions that are comparable with the reference run. Even in case of absolutely ignoring the OWF for more than 44 h, that is, after 1.8 days, the ocean's response to the OWF does not fully disappear. Comparing the three operation cases, it can be said that the stronger and longer the OWF acts on the atmosphere and the ocean, the longer and stronger the ocean is affected.

Whereas changes in hydrographic (Fig. 5.21) do not end up in surprising physical differences, the velocity field (Fig. 5.20) leads to horizontal circulation, which strongly affects the vertical component.

Foremost, the *velocities at surface* are analyzed. As expected, with the turning on of the OWF and with it an increase of wind wake, the ocean velocity field is affected by speed reduction in the wake area. Maximal decrease is reached till the point of turning off the OWF. Although the first operation time is quite short, with 4.2 h for opc01 & opc02 and 2.6 h for opc03, the reduction counts around -0.1 m/s for opc01 & opc02 and 0.07 m/s for opc03. As indicated in Sect. 5.1, the v-component of velocity encroaches into the dynamical system by strong changes close to the OWF. The change of v-component, compared to the reference run, is dominant at the end of the first OWF operation duration.

The v-component increases by 0.04 up to 0.07 m/s at the surface. It is a horizontal effect to counteract against the wake in u-component and thus the produced dipole of ζ. At the point of turbine shut down, the u-component increases again, while the v-component is reduced, additionally, w-component is changed to keep the equation of motion.

During that time, the dipole structure of *surface elevation* is built on factual connection—the stronger the wake is, the stronger the tilt of ζ is. The formation of extrema for ζ is delayed, compared to the velocity wake, by more than 5 h for all three cases. Furthermore, it is recognized that surface elevation ζ does not consequently decrease by time during the 'off' phase of the OWF. It decreases by pulsing.

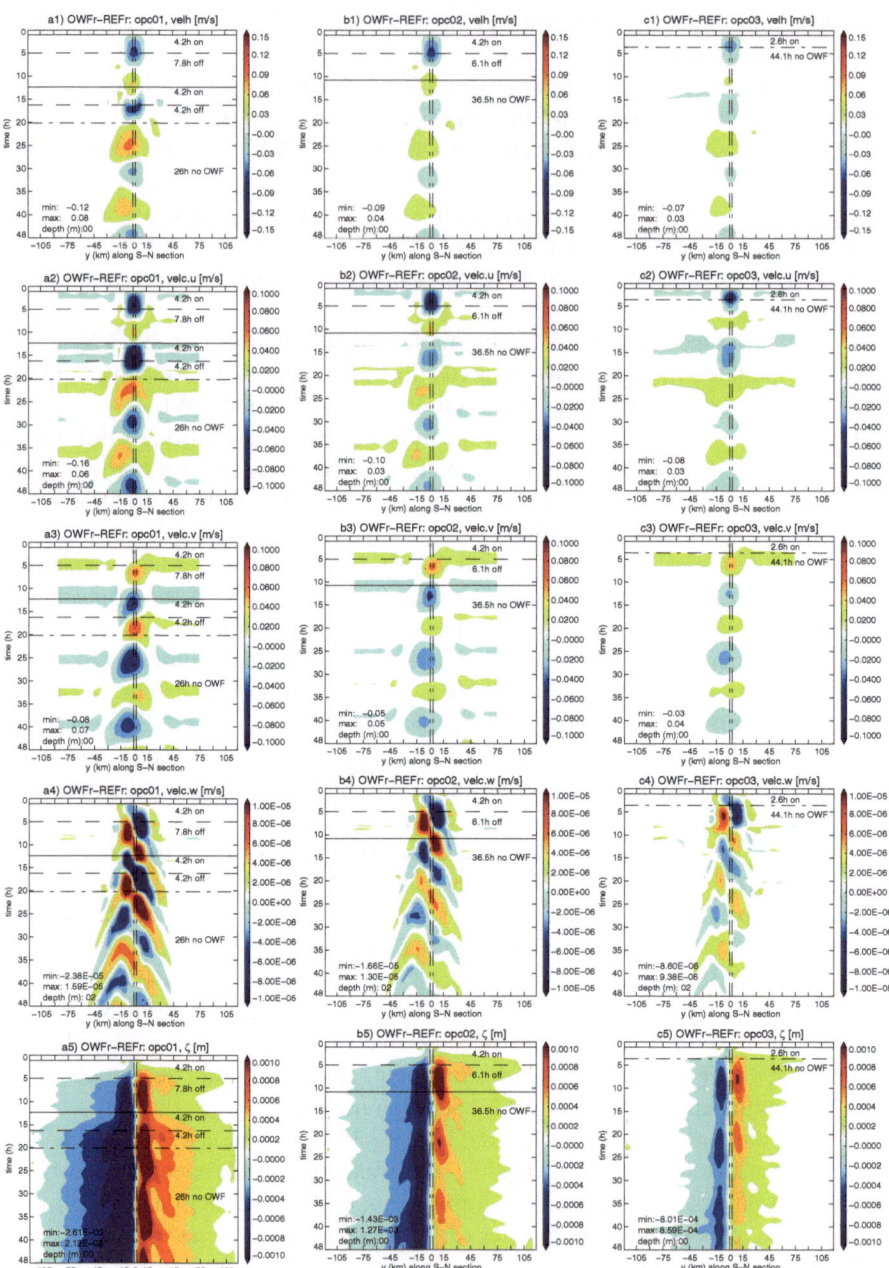

Fig. 5.20 Hovmöller diagrams of OWF effect (OWFr–REFr) of hydrodynamics depending on three different cases of OWF operation opc01 (**a1**–**a5**), opc02 (**b1**–**b5**), and opc03 (**c1**–**c5**) along the S–N cross section. Illustrations (**a1**)–(**c1**) show results of horizontal velocity field at the surface, (**a2**)–(**c2**) of velocity component u at the surface, (**a3**)–(**c3**) of velocity component v at the surface, (**a4**)–(**c4**) of velocity component w at 2-m depths, and (**a5**)–(**c5**) of surface elevation. *Horizontal black lines* clarify mode of OWF operation. Here, the *solid lines* show the start of OWF

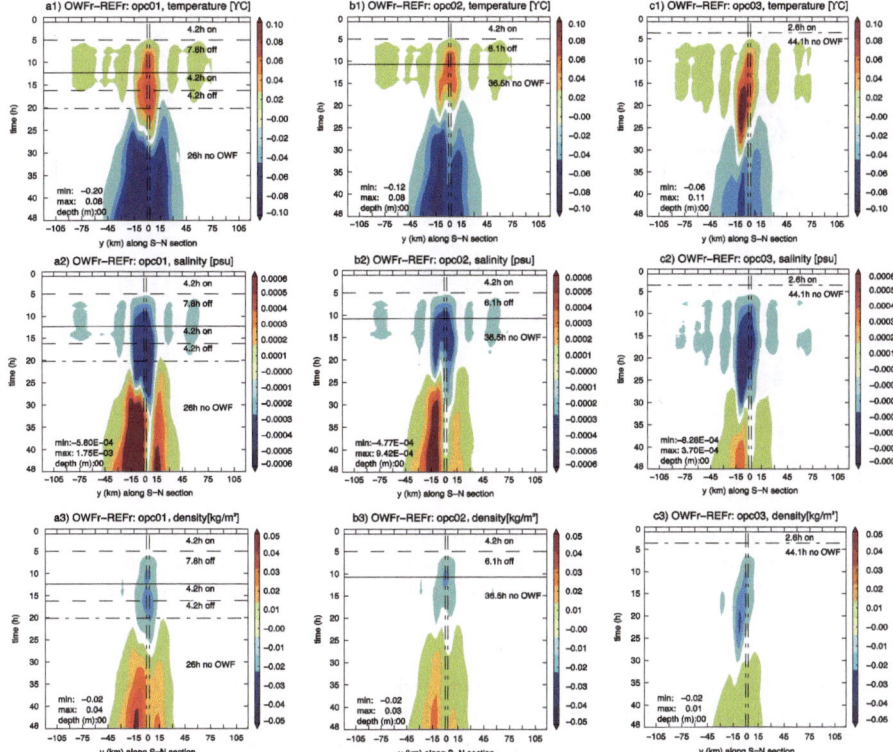

Fig. 5.21 Hovmöller diagrams of OWF effect (OWFr–REFr) of hydrographic depending on three different cases of OWF operation opc01 (**a1**–**a3**), opc02 (**b1**–**b3**), and opc03 (**c1**–**c3**) along the S–N cross section. Illustration (**a1**)–(**c1**) show results of SST, (**a2**)–(**c2**) of salinity at the surface, and (**a3**)–(**c3**) of density at the surface. *Horizontal black lines* clarify mode of OWF operation. Here, the *solid lines* show the start of OWF operation mode, while the *dashed horizontal lines* stand for switching off/ignoring the OWF. The OWF district is around $y = 0$ km and counts 12 wind turbines. Maximal changes occur parallel to extreme changes of surface elevation

In a time difference of 14 h, a weaker extrema of ζ exist, weakened over 7 more hours, followed by an increase again to a little weaker extrema than before. Extrema are placed close to the OWF, but depending on the first 'on' duration, the horizontal is differently affected in the three operation cases. The pulsing of ζ is connected with the vertical cells. Changes in surface elevation and velocity field are concentrated at the OWF region; only surface elevation affects more horizontal area by time.

<hr>

Fig. 5.20 (continued) operation mode, while *dashed horizontal lines* stand for switching off/ignoring the OWF. The OWF district is around $y = 0$ km and counts 12 wind turbines. While horizontal velocity component u acts with wind forcing, maximum changes of the residual dynamical variables occur time shifted. After 1.8 days without OWF signal, the OWF signal on the ocean still exists, albeit weaker

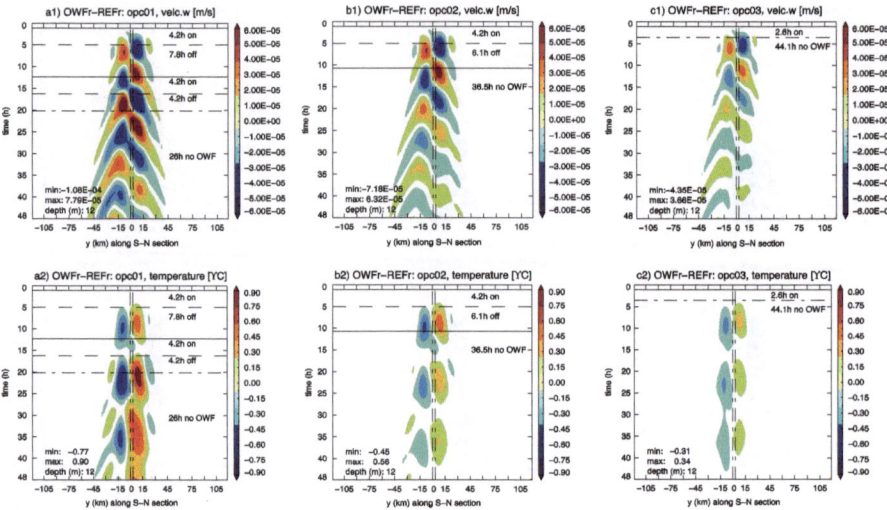

Fig. 5.22 Hovmöller diagrams of OWF effect (OWFr–REFr) in 12-m depths of vertical velocity component w (**a1–c1**) and temperature (**a2–c2**) along the S–N section for the three OWF operation cases opc01 (**a1–a2**), opc02 (**b1–b2**), and opc03 (**c1–c2**). Inertial oscillation is obviously seen at w-component

As mentioned, the strongest difference, compared to theoretical run, with constant forcing over time is detected at vertical *velocity component w* (Fig. 5.20). One can expect that turning on an off wind turbines will lead to variation in the horizontal velocity field which affect surface elevation and the vertical motion. But the dynamic is not only pulsing; related to OWF operation, an additional side effect occurs.

Shifted by time, the velocity component relatively strongly increases by turning off the OWF after 4.2 h in all three operation cases.

The diversity between increase and decrease of horizontal velocity components due to the OWF operation ends in a pulsing of vertical cells, which rotates counter clockwise around the OWF. With time, the core of upwelling/downwelling rotates around the OWF, which leads to the alternating trend with time. The rotating effect only strongly affects an area ±30 km around the OWF center and affects all depths. Such alternating of velocity triggers the increase and decrease of the surface elevation. The rotation of the vertical cells occurs due to a provoked inertial oscillation by turning on and off the wind turbines. The movement of vertical cells is counterclockwise, with a period of 13–15 h, which agrees with a mean inertial oscillation T around the 55.00° latitude of 14 h based on

$$T = \frac{2\pi}{2\Omega \sin \varphi} \tag{5.8}$$

Due to no coasts and no tides in the model ocean box, the inertial oscillation cannot be suppressed like in nature.

In the *hydrographic fields*, the alternating effect cannot be observed at the surface; see Fig. 5.21. Here, it is clearly seen that the OWF effect on the ocean's SST acts delayed, becoming obvious after turning off the OWF. The OWF leads to a stronger and longer increase of SST than in the operation cases opc01 and opc02 after the first power on period. Here, the cooling occurs around 5 h earlier due to stronger vertical velocities supported by stronger start wake and switching on turbines one more time.

Operation case opc02, where the OWF is turned on only once, ends faster in cooling but is not as strong as opc01, Fig. 5.21. In the case of opc01, where the OWF is turned on again after 7 h, an SST warming is longer kept within the OWF influenced area. Similarities occur for salinity and density; see Fig. 5.21. The rotation of vertical cells around the OWF affects temperature not at the surface but in the ocean depth. Figure 5.22 depicts the evolution of the vertical velocity component w and the temperature at the thermocline in 12-m depth. With a delay in the vertical velocity component, a significant temperature reaction occurs, but the warming/cooling does not rotate around the OWF like the cells of vertical motion. The warming (north of the OWF) and the cooling (south of OWF) only vary due to the rotation of the vertical cells. Due to the inertial oscillation, the effect on temperature can be nearly reduced at the turning point from positive vertical motion to negative vertical motion (Fig. 5.22), but the effect on the temperature field is strengthened again when the downwelling/upwelling cell takes effect in north/south of the OWF.

The inertial oscillation avoids a quick reestablishment of starting conditions. It also shows how sensible ocean dynamics are related to a wind field. Hence, even a short operation of OWF can induce a mixing, which is connected with temperature changes of 0.5 °C up to 1.0 °C within a couple of hours.

5.3.2   *Analyzing OWF Effect on the Ocean Depending on Wind Speed*

The OWF induced wind wake depends on, besides size and power of the wind farm, the wind speed. Stronger wind speeds result in different strength of wind wakes behind the wind farm, which again leads to variation on the effect on the ocean.

To evaluate the wake effect on the ocean related to wind speeds, three wind speed cases were analyzed. So far, only the wake effect in case of one wind speed based on ug = 8 m/s has been discussed. In the case of METRAS, wind turbines operate in case of wind speeds at hub height of 2.5 up to 17.0 m/s (personal correspondence with M. Linde). Thus, the sorts of wind speeds for analysis comprises ug = 5 m/s (UG5), ug = 8 m/s (UG8), and ug = 16 m/s (UG16). The wind forcing shows that in all three cases, the affected area is similar, but stronger wind speed leads to a more intensified effect, Sect. 5.2.2.

Simulations are based on TOS-01 with the wind farm of 12 turbines directly impacting an area of 36 km^2. Corresponding simulations for result presentation are

T012ug05 TS01HD60F01, T012ug08 TS01HD60F01, and *T012ug16 TS01HD60F01.*

Because of the similar structure of effect on ocean, this analysis concentrates on a single-point comparison. The chosen points mark the middle of the model area, the wind farm itself, as well as areas of extreme changes due to the OWF. In sum, 11 positions at different levels, which are surface, 12 m (depth of the thermocline), 30 m (half of model depths), and 60 m (bottom), are included in this analysis. These positions are pictured in Fig. 5.23.

A single point analysis is chosen to accent discrepancies over the whole OWF-affected area. Independent of wind speed velocity wake, dipole in surface

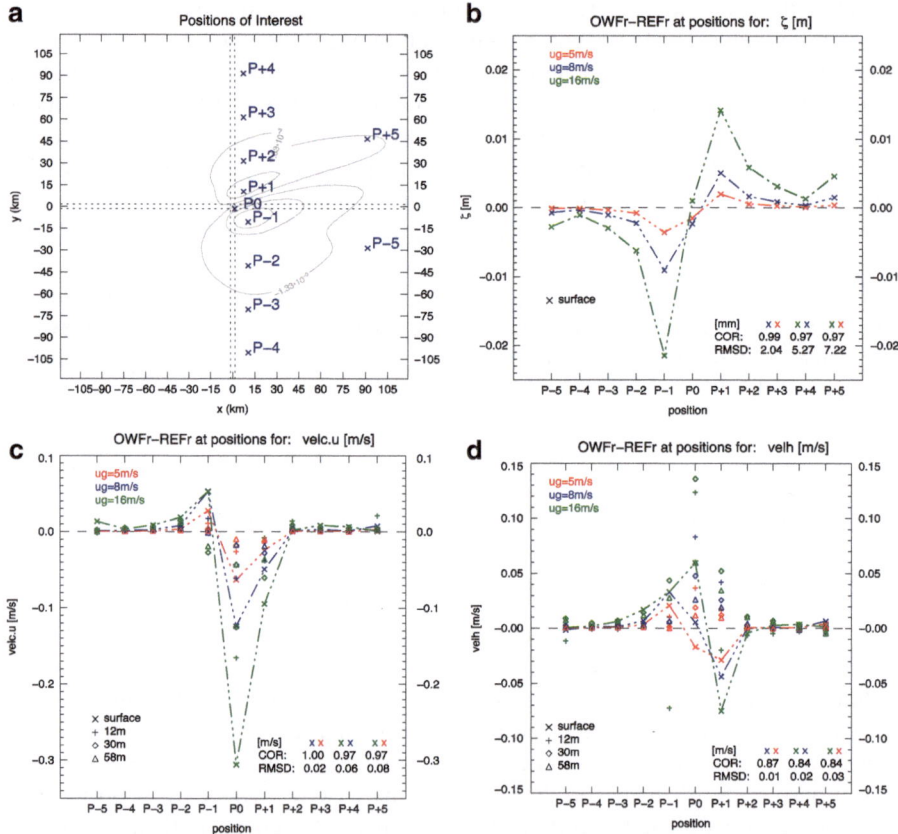

Fig. 5.23 (a) Positions of interest for analysis of different wind speed conditions chosen by effect's extreme (here clarified by *gray contour lines* of surface elevation in *m*) *x*-position. *P0* points the OWF center, *P+* is 6 km, *P−* 9 km easterly. Illustrations (**b**)–(**c**) show OWF's effect after 1 day of simulation by surface elevation *ζ* (**b**), velocity component *u* (**c**), and horizontal velocity field (**d**) forced by ug = 5 m/s (*red*), ug = 8 m/s (*blue*), and ug = 16 m/s (*green*). Different symbols stand for different depths: '×' surface, '+' 12 m, '◇' 30 m, and '△' 58 m close to bottom. COR gives correlation and RMSD the root mean square for corresponding values and depths

elevation and vertical cells occur. In case of UG 5/8/16 m/s, vertical cells have a dimension of 15/30/30–40 km width and clearly affect depths till 30/60/60 m. Therefore, difference in simulations at investigation positions dominantly varies between $P - 3$ and $P + 3$ for all variables.

Correlation (COR) of results along the single points and the root mean square difference (RMSD) are used as a statistical tool to analyze accuracy of simulation results based on forcing of three different wind speeds. Here, RMSD is used as a measure of the discrepancy among the three different model samples to compare values due to different forcing cases. In the following, the effect on the ocean due to the wind farm is examined.

All the analyses of the three sets of wind speed forcing in Sect. 5.2.2 show that a relatively stronger wind field means already a more intense wind wake. Considering METRAS' wind turbine parameterization, the rotor thrust grows with the cube of velocity, so in the case of higher wind speeds, turbines are able to detract more energy out of the atmosphere and therefore wind reduction behind OWF is intensified. An intense wind wake results in an intense velocity wake of the ocean velocity u-component. Figure 5.23 documents that fact. The overall extrema of variables mostly increase with increasing wind speeds. The strongest effect is identified at the thermocline at 12-m depth, while at the surface, 30 m, or at the bottom, effects are consequently smaller.

The effect of *velocity component u* along investigation points $P - 5$ to $P + 5$ over the area, based on different UG speeds, has its wake at position $P0$ within the OWF and is highly correlated by 1.00 between forcing UG5 and UG8 and 0.97 between UG5, respectively UG8, and UG16. By means of the correlation of the three different *horizontal velocity fields*, it becomes clear that velocity component v exerts a diverse impact on the velocity field. The stronger is the wind forcing, the higher are the discrepancies. Correlations along investigation points with UG16 are only of 0.84 at the surface. Between UG5 and UG8 forcing exist little fewer differences; 0.87 correlates them. Wake magnitudes at the surface have a linear character, and RMSD increases with wind speed difference.

The effect of *surface elevation* ζ shows a similar distribution, having a maximum at $P + 1$ and a minimum at $P - 1$, with a high correlation between UG5, UG8, and UG16; see Fig. 5.23. Results for ζ at investigation points have a weak root mean square difference between the runs of different wind speeds. The stronger is the wind forcing, the more distinctive is the extreme of ζ dipole structure. The growth of extrema leads to an exponential character but cannot be specified due to only three wind cases.

The fact that a stronger wind results in a stronger effect on the ocean surface, the effects on vertical velocity and temperature must be consequently intensified. The effects on *vertical velocity component* and *hydrographic parameter*, compared by different forced wind speeds, are pictured in Fig. 5.24. The maximal upwelling along investigation points occurs at point $P - 1$ for all three wind cases through all layers. The strongest downwelling is placed at $P + 1$ for wind case UG5 and UG8, while for UG16, the maximal downwelling can be found at $P0$.

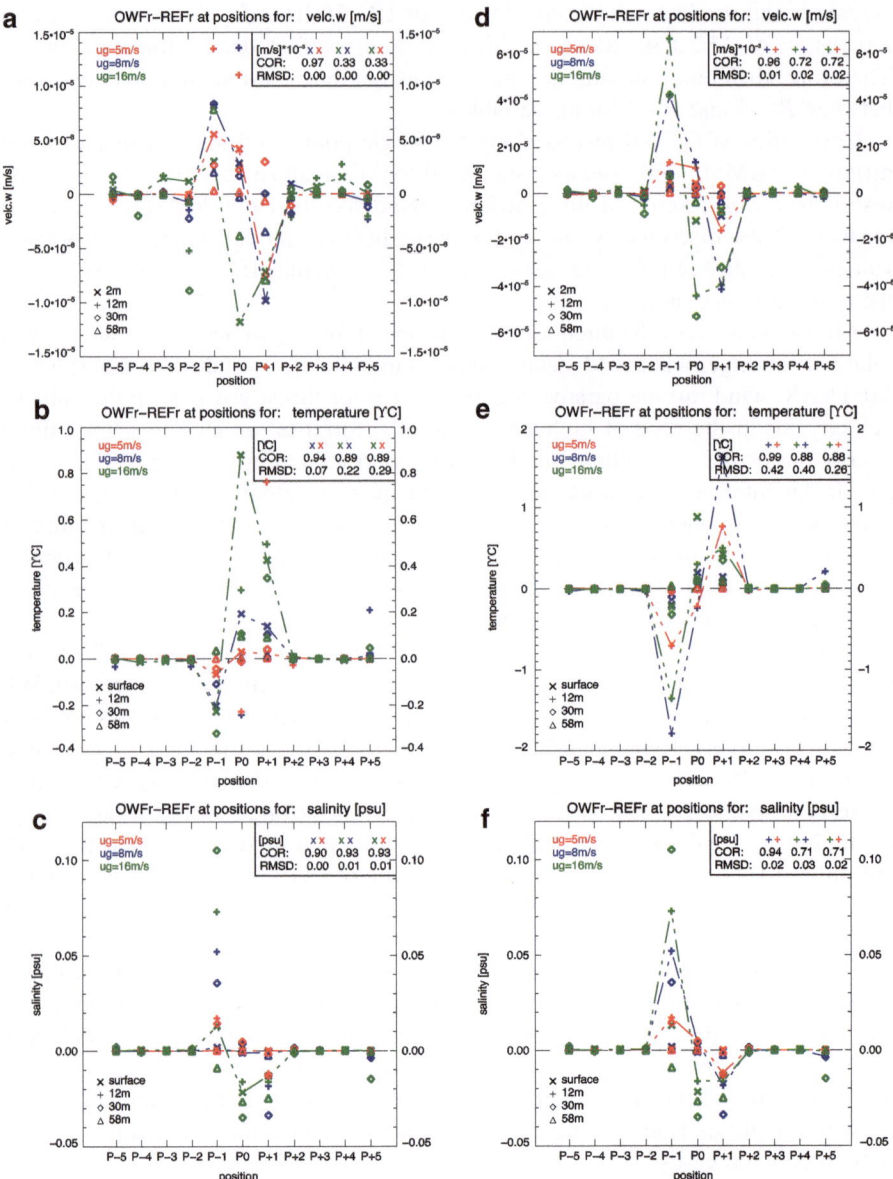

Fig. 5.24 OWF's effect on the ocean due to three different wind forcing after 1 day of simulation at positions, given in Fig. 5.23a. Illustration (**a**)–(**c**) focuses on the surface, respectively 2 m for velocity component, (**d**)–(**f**) on 12-m depth. Rows show from top to bottom (**a**), (**d**) velocity component w; (**b**), (**e**) temperature; and (**c**), (**f**) salinity forced by ug = 5 m/s (*red*), ug = 8 m/s (*blue*), and ug = 16 m/s (*green*). $P0$ points the OWF center, $P+$ is 6 km, $P-$ 9 km easterly of the N–S cross section. Different symbols stand for different depths; '×' surface (respectively 2 m for velc.w), '+' 12 m, '◇'30 m, and '△' 58 m close to bottom. CORR gives correlation and RMSD the root mean square for corresponding depths marked with (**a**)–(**c**) symbol '×' and (**d**)–(**f**) symbol '+'

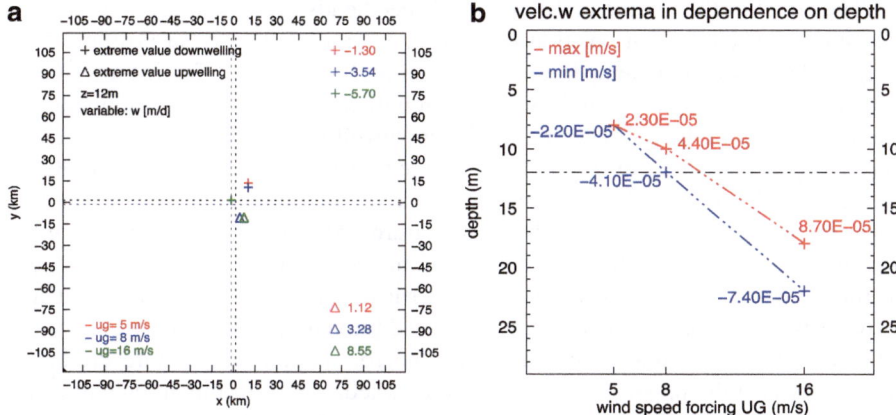

Fig. 5.25 Position of (**a**) extremes of variable w in 12-m depth and (**b**) extremes of w in the vertical along the y-section based on different wind speed forcing. In the depth of the thermocline, that is, 12 m, the position of positive extreme is comparable between different wind speeds, while the negative extreme swifts into the OWF with increased wind forcing. Negative maxima are slightly higher than the positive ones and are positioned in deeper layers

Figure 5.25 illustrates that maximal upwelling velocity at 12 m nearly fits the same position; only the run based on UG16 forcing has its maximal upwelling one grid box (~3 km) easterly of the result based on UG5 and UG8 forcing. Regarding maximal downwelling, the effect at 12 m of run with UG16 forcing has its extreme within the OWF, while extremes of UG5 and UG8 forced runs are placed more northeasterly. It seems that with increased wind forcing, the position of downwelling switches more towards midpoint of the OWF, but only based on three cases, it is not possible to make a statistical fundamental statement. Nevertheless, it can be underlined that the stronger the wind speed forcing is the stronger the vertical motion will be, especially in deeper layers. The strongest upwelling at 2 m, at position $P - 1$, is identified by simulation with forcing UG8 (Fig. 5.24a). This does not enervate the termed relation because at the surface, extrema are also affected by horizontal flow, and so positions of extrema are not defined at the same location, which also causes a bad correlation. That is why here run, forced with UG16, gives only the lowest upwelling. However, maxima/minima of vertical velocity w is $1.04 \times 10^{-5}/-0.77 \times 10^{-5}$ m/s (for forcing UG5), $1.54 \times 10^{-5}/-1.10 \times 10^{-5}$ m/s (for UG8), and $0.0246/-0.0194$ mm/s (for UG16) over the whole area and within first 2-m depths.

Further, an increase in wind speed forcing results in a swift of positions of maximal and minimal vertical velocities into higher depths. Figure 5.25 clarifies the dependence between induced wind speed, depth, and extrema. Additionally, it can be said that the upwelling effect is a little bit more dominant than the downwelling effect and that the discrepancy between these two motions increases with wind speed.

The effect of the forcing of 16 m/s wind speed ends in a *temperature* increase at the surface of 0.88 °C at position $P0$ and a decrease of 0.23 °C at position $P - 1$ (Fig. 5.24b). Maximal temperature decrease at $P - 1$ due to forcing of UG8 is slightly weaker, -0.20 °C, and for UG5 forcing, only -0.07 °C. The temperatures are higher than in the reference run at the downwelling area. This forcing of UG16 has the strongest effect of 0.88 °C increase, compared to 0.19/0.03 °C for UG8/5. Although the vertical velocity component w at the thermocline is more intense for UG16 forcing than for UG8, the temperature extrema at the thermocline are maximal in the case of forcing UG8. This underlines the previous analysis that vertical mixing is not singly driven by vertical velocity component w but also driven by additional exchange processes. Additionally, a stronger wind forcing quickens processes; that is also a reason the for weaker temperature effects by UG16 because the upper layers are quickly mixed, which results in one abundant upper layer. In turn, the thermocline is pushed into deeper depths.

The shift of the *thermocline's* position is pictured in Fig. 5.26. The change from UG5 to UG16 forcing at thermocline is clearly seen. Excursion increases with wind speed forcing from 2 (UG5) over 4 (UG8) up to 8 (UG16) meters. In the case of UG16, the thermocline is switched beginning from 12 to 20-m depths. Correlations of temperatures with forcing UG16 are around 0.15 higher than for the vertical velocity component w at the thermocline. Correlations between UG5 and UG8 are high with 0.94 at the surface and 0.99 at the thermocline.

The effect on *salinity* forced by the three different wind speeds has a weak variability in comparison due to goof correlations of around 0.9 at the surface, Fig. 5.24c, f. Differences of the effect mostly occur in the downwelling region at positions $P0$ and $P + 1$, especially in the thermocline depth. However, the OWF effect on salinity increases with wind speed forcing.

Upshot

Summarizing, the effect of an OWF on the ocean, considering different wind speeds, does not impact strongly on horizontal dimensions. Dipole structure and up- and downwelling cells have similar dimension because they depend more on the OWF arrangement. Positions of maxima and minima vary due to differences in the horizontal velocity field, which depends on the wake intensity. Here, a trend of extrema moving towards the OWF grid boxes with higher wind speed forcing was detected. That behavior is based on the fact that the maximum decrease in wind is placed close to the OWF and so the effects here are stronger on the ocean. The cells of vertical motion become more intense with higher wind speeds and the cells' extrema occur in deeper layers, which more strongly affects vertical layers again. Stronger OWF induced wind wakes support the vertical mixing and lead to a stronger exchange of temperature via the thermocline. Hence, the depth of the thermocline increases with wake intensity. The variation of the OWF effect due to different wake intensity in temperature is in order of tenths of a degree, even for the horizontal velocity component u, and variations of the surface elevations count several millimeters; for the vertical motion, hundredths of mm/s.

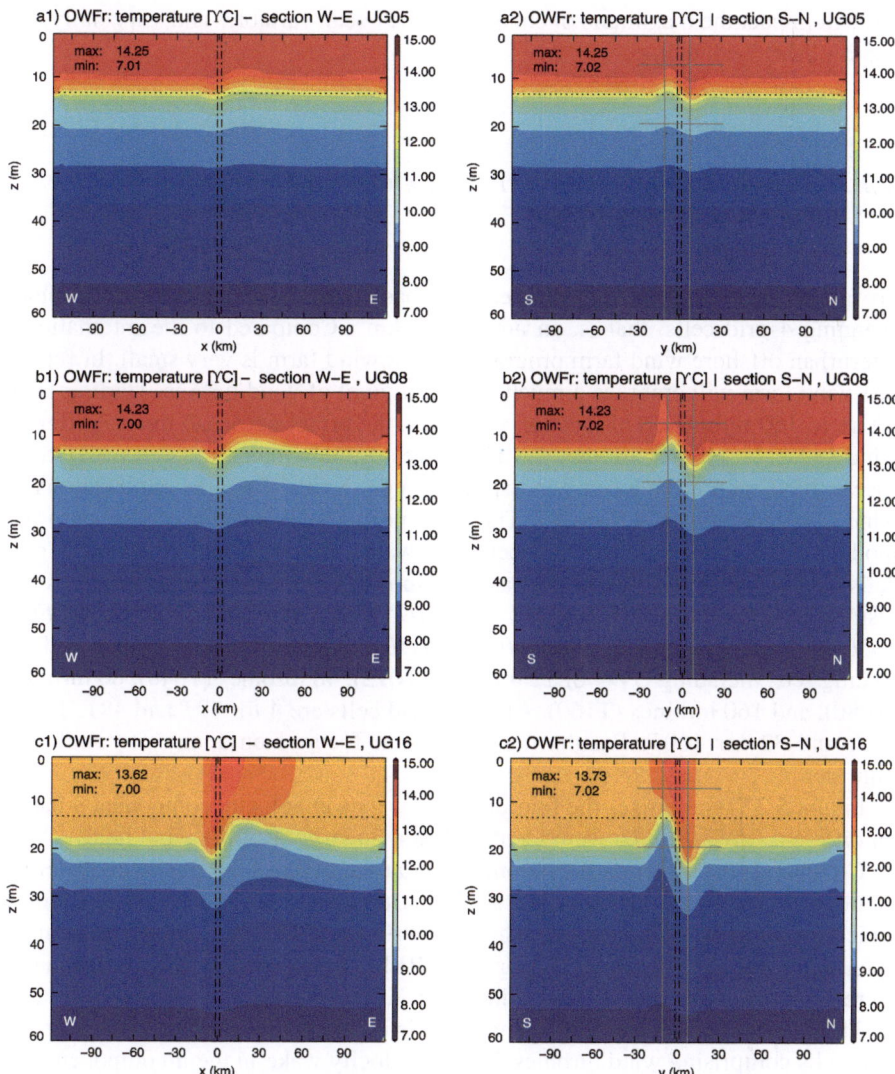

Fig. 5.26 Temperature profiles along the (**a1**)–(**c1**) W–E and (**a2**–**c2**) S–N sections through the OWF based on different wind speed forcing of (**a1**)–(**a2**) 5 m/s (UG5), (**b1**)–(**b2**) 8 m/s (UG8), and (**c1**)–(**c2**) 16 m/s (UG16). Excursion of the thermocline around the OWF based on UG08 is marked by *solid black horizontal lines*. Excursion increases with stronger wind speed. Additional, thermocline is shifted to deeper layers

Overall, an increase in wind speed leads to an intensification of the OWF effect and accelerates the ocean's reaction to the wind wake in order of tenths and hundredths. If we consider that shown results after 1 day of simulation are based on runs forced by a constant wind field, we can assume that slight changes over a day in the wind field will not strongly affect the ocean's reaction on the wind wake but will impact the magnitude of changes on the ocean variables and, especially, the depth of the thermocline.

5.3.3   Analyzing OWF Effect on the Ocean Depending on Wind Park Power

So far, presented effects on the ocean are based on a wind farm of 12 turbines spanning 4 grid cells, that is, an area of 36 km^2. Compared to the international interurban offshore wind farm program, such a wind farm is very small. In future, one wind farm will embrace a much higher amount of wind turbines starting from 50 up to 160 turbines. This section illustrates the effect on ocean due to different wind wakes based on various wind farm sizes. Simulations are based on TOS-01 (ocean box) with an induced wind speed of ug = 8 m/s. Detailed explanations of wind forcing are given in Sect. 4.2. There it is perceived that due to model setup and model resolution, distinctions between wind wakes are rare and that due to big grid cells of 3×3 km, the affected areas are similar. Corresponding simulations for result presentation are *T012ug08 TS01HD60F01, T048ug08 TS01HD60F01, T080ug08 TS01HD60F01, T160ug08 TS01HD60F01.* Simulations are based on forcing sets, including OWF of 12 turbines (T012), 48 turbines (T048), 80 turbines (T080), and 160 turbines (T160). Affected grid cells are 4 for 12T and 48T, 16 for 80T, and 32 for 160T. Results here are again focused on 1 day of simulation time step.

Figure 5.27 summarizes the four different wakes in velocity component *u* based on different OWFs, which were identified as a fingerprint of the wind field.

Comparing the effect on the *u*-component at the surface, the runs with different numbers of wind turbines leads to a similar structure of the OWF induced change, which mostly depends on the OWF area. In a run with 80 wind turbines, a stronger exhaustive change between OWFr and REFr occurs, especially a northerly increased wake flank (Fig. 5.27c1), due to stronger wake flanks in the wind forcing.

The affected area of runs T012 and T048 is identical due to the same number of grid cells comprising wind turbines, but the velocity wake in the *u*-component is a little more intense by T048 with a minimum at the surface of −0.139 m/s, compared to −0.136 m/s for T012. A reason for the difference rests in a different magnitude of the wind field due to the considered power and amount of wind turbines in METRAS wind turbine parameterization. In the case of T080 and T160, the wake is deformed by the OWF area, and therefore the wake width in *y*-direction is bigger for T080 and T160, compared to T012 and T048. As in case of the wind field, the ocean's *u*-wake grows with the number of wind turbines, which means with the OWF area. Depending on the affected area by the wind wake, the *u*-component and already the velocity field implicate a disturbance that is more addicted to the wind wake district. In all four cases, the minimal *u*-velocities occur within the OWF or, in case of a wider OWF, more northerly. Wake flanks are more intense in case of a

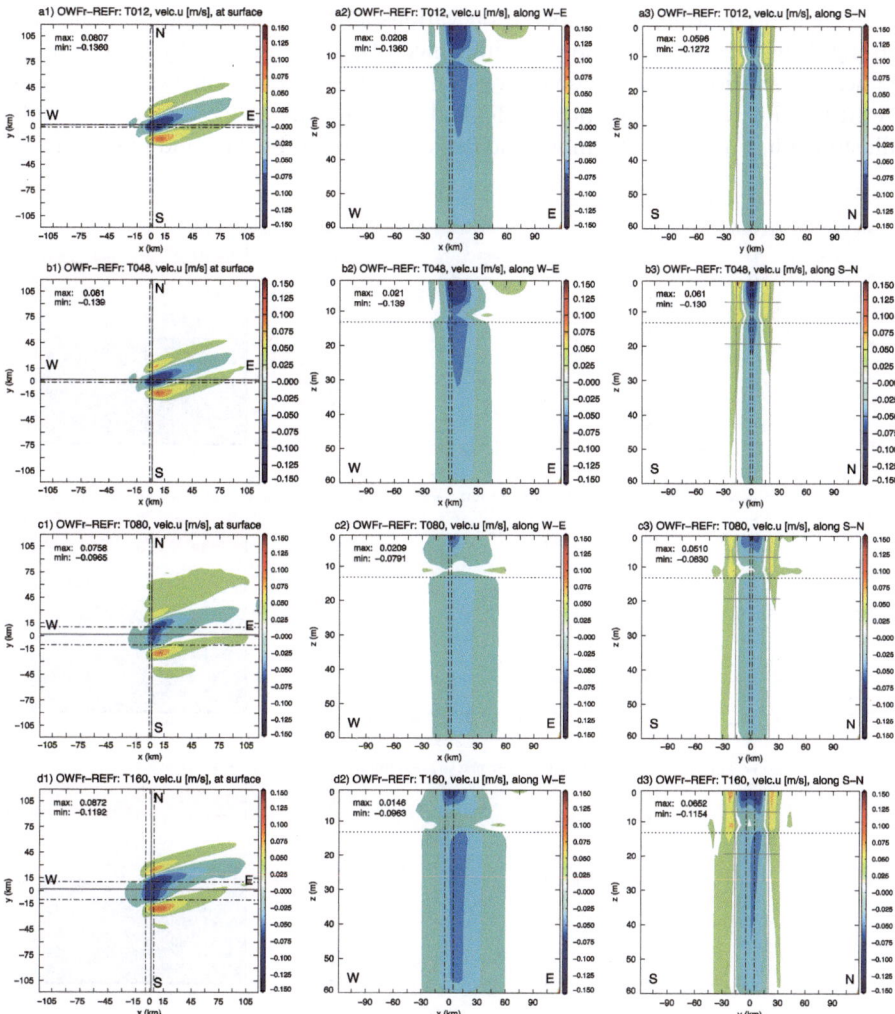

Fig. 5.27 Change of velocity component u (velc.u) due to different amounts of wind turbines. Images (**a1**)–(**a3**) illustrate velocity component velc.u in the case of T012, (**b1**)–(**b3**) T048, (**c1**)–(**c3**) T080, (**d1**)–(**d3**) T0160 at (**a1**)–(**d1**) surface, (**a2**)–(**d2**) along the W–E section and (**a3**)–(**d3**) along the S–N section through the OWF. *Dashed–dotted lines* encase the OWF district. *Solid lines* denote the cross section through the OWF, and *dotted lines* in section plots mark the depth of the thermocline based on T012. Units are given in m/s

bigger OWF, and so the wake flanks of increased velocities more strongly impact deeper layers than T012.

The maximal wake magnitude is given by T048, as mentioned in Sect. 4.2; the reason for that context is that in T080 and T160, more grid cells, filled with fewer wind turbines, are astonished by the wind turbines.

Although the wind wakes, based on various amounts of turbines, do not show strong difference in magnitude, the influence of more grid cells by wind turbines indicates variations in ocean conditions. In all four cases, the changes occur in the vertical, with two main vertical cells, and consequently in the temperature field, illustrated in Fig. 5.28. The more grid cells are affected by the wind reduction, the

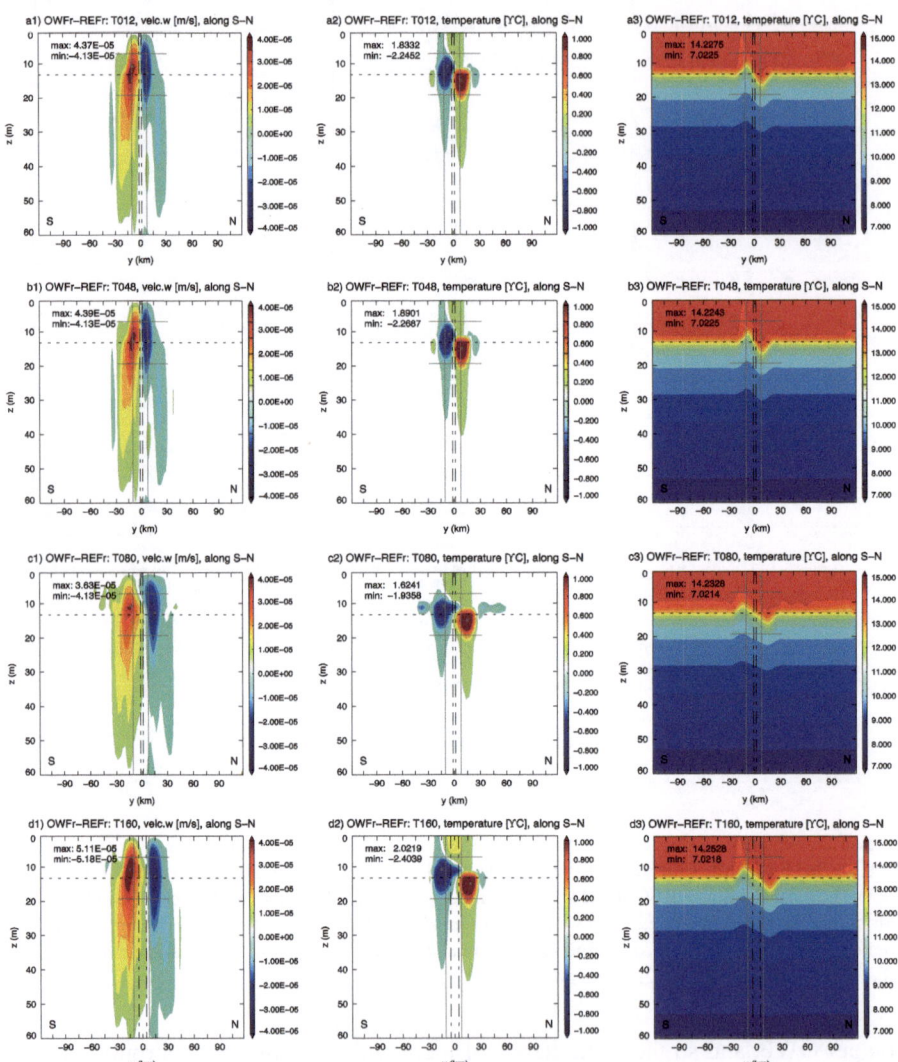

Fig. 5.28 Effect of (**a1**)–(**d1**) vertical velocity component *w* and (**a2**)–(**d2**) temperature and (**a3**)–(**d3**) temperature stratification of the OWF run based on 12-turbine (**a**), 48-turbine (**b**), 80-turbines, (**c**) and 160-turbine (**d**) OWF along the *y*-section through the OWF from S to N. Crosses mark the position of maxima and minima, *dashed line* illustrates the depth of thermocline, and the limits of thermocline exclusion is marked by short *black solid lines* based on T012

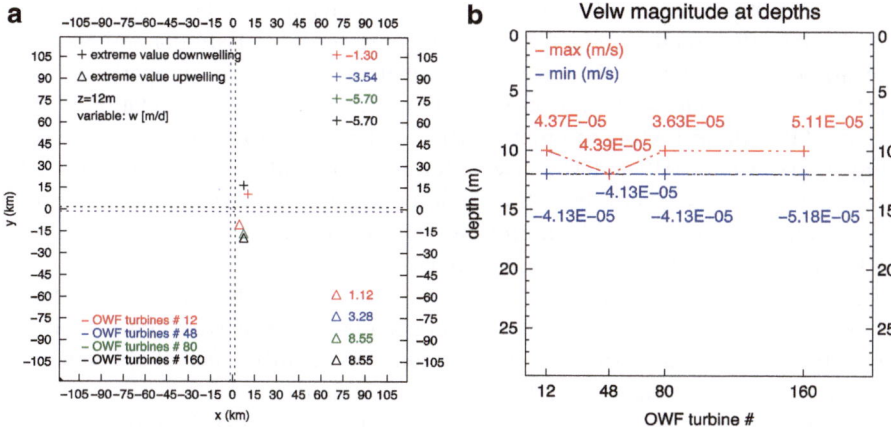

Fig. 5.29 (a) Location of extremes of variable w in 12-m depth and (b) variations of overall maxima and minima along the y-section from S to N in dependence of depth

more intense are the vertical velocity cells with a maximal global mean of downwelling and upwelling for T160 of $-1.68E-06$ and $+1.83E-06$ m/s. The extrema along the S–N cross section through the OWF are in order of 10^{-5} m/s, and the bigger is the OWF, the stronger is the intensification of the vertical motion, especially for downwelling. While upwelling varies stronger between the simulations, the downwelling is constant with a minima of $-4.13E-05$ m/s.

The vertical effect is also quite consistent. The extrema of velocity component w in 12-m depth have the same maxima/minima location for T012 and T048, respectively T080 and T160 (Fig. 5.29a). The extrema of the vertical cells along the S–N section are also located in the same depths for all turbine assumption, apart from T48, which has a maximal upwelling in 12 m and not in 10 m (Fig. 5.29b). A maximal downwelling is registered along the S–N cross section in 12-m depth for all OWF sizes.

Figure 5.28 also represents the effect of wind turbine amount on the temperature in the vertical. Thermocline stays at 12-m depths for all four simulations. In space of intense vertical mixing, temperature variations occur with maxima beneath and minima above the thermocline.

Among run T080, the temperature stratification is only affected around the OWF. Simulation run T080 shows a little cooling at the thermocline along the S–N cross section due to the different wind fields and therefore a diffuser surface elevation. It can be said that a different number of operating turbines affects the excursion of the thermocline around the OWF. The excursion of the thermocline is more horizontally distorted in the case of greater OWF districts (Fig. 5.28a3–d3). But the horizontal distortion has hardly influenced the vertical dimension of excursion in contrast to wind speed effects. Neglecting T080, the change in the temperature increases with the OWF sizes due to temperature advection by vertical motion and is supported by wider vertical cells of changes and horizontal diffusion.

Figure 5.30 clarifies the fact that affected grid cells play a much more important role for the effect on the ocean than the number of turbines. Extrema of the surface elevation increase with OWF size and number of turbines. A comparison of the surface elevation along the S–N cross section results in a good correlation of 0.90 up to 1.00 with an RMSD in order of 10^{-3} and 10^{-4} m for runs T012 and T048 (Fig. 5.30a). Runs T012 and T048 (wind turbines are scattered over four grid cells) show along y-cross-section from S to N at the surface, as well as in 12-m depth a high correlation of 0.99 and better for the velocity components u, v, and w, as well as for hydrographic variables (Fig. 5.30b–f).

Independent of v-component, T080 and T160 are even well correlated for almost all ocean variables by 0.99 and better. Discrepancies between T080 and T160 in the velocity v-component (correlation of 0.77) relate to the difference of affected grids (16 and 32) and discrepancies in the wind forcing field.

The important factor of the OWF size regarding grid cells is highlighted by the comparison of runs with OWFs over four grid cells (T012&T048) with runs with OWFs over more grid cells (T080&T160). Here, correlations, along representative S–N cross section, are almost bad for velocity components u and v and tend often to the statement of not being correlated. Horizontal variations due to the OWFs along the cross-section S–N impact the correlation value in Fig. 5.30a–b. The simulations of different OWFs result in good correlations for the variables velocity component w, temperature, and salinity at 12-m depths along cross-section S–N (Fig. 5.30d–f). Here, the stronger agreement underlines the connection of formation of up- and downwelling to surface elevation. The change of the ζ dipoles more to the north for ζ increase and to the south for ζ decrease is also identified in w-component and hydrographic conditions. The distortion of the thermocline in dependence to the OWF area is also documented in Fig. 5.30e based on the horizontal dimension of the temperature excursion.

Upshot
The comparison of the effect on the ocean due to different wind farm sizes results in the assumption that the wind farm's size does not impact vertical mixing in a direct way. Different wind speed forcing cases more strongly impact on the vertical stratification. But the wind farm size amplifies the effect in the horizontal. The magnitude of vertical mixing slightly varies due to the different amount of affected grid cells, and the maximal anomaly between T012 and T160 in the temperature counts 0.25 °C. The results of that analysis that greater OWFs have a comparable effect on the ocean to smaller wind farms come in force. But based on issues of the horizontal resolution and wind wake presentation, it must be considered that a finer grid in the case of T012 may result in a horizontally smaller wake dimension orthogonal to the wind direction. Hence, the effect on ocean would be weaker.

Therefore, the outcome of that analysis cannot be generalized at this point. A more detailed wake illustration simulated by higher resolution would be necessary for a final statement.

But results can be generalized for velocity wakes at ocean surface. A slightly wider wake does not change stratification in the vertical but triggers impact in the

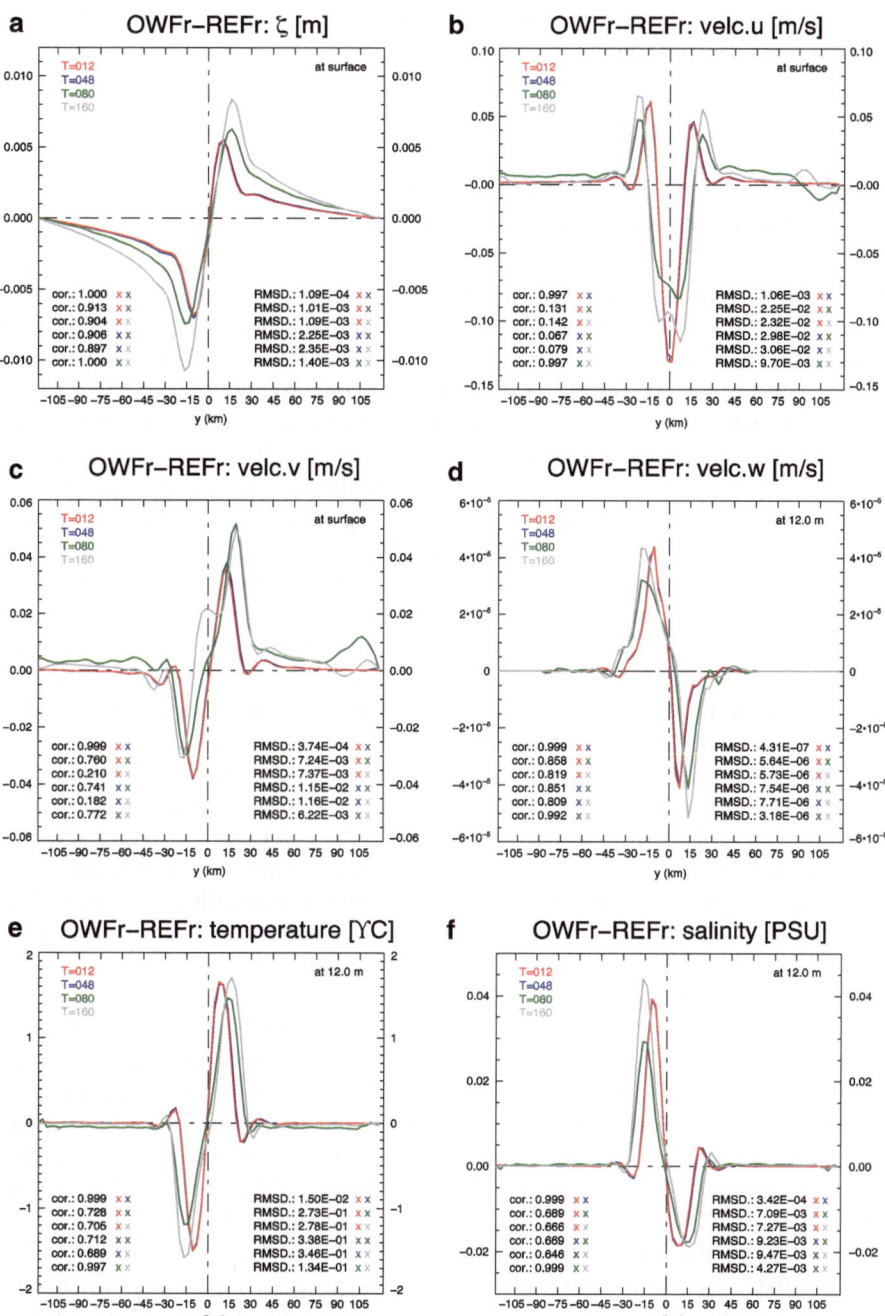

Fig. 5.30 Comparison of ocean variable changes due to OWFs: (**a**) surface elevation, (**b/c**) velocity component *u/v* at the surface, (**d**) velocity component *w* at 12 m, (**e**) temperature at 12 m, and (**f**) salinity at 12 m. Results are colored for T012 in *red*, T048 in *blue*, T080 in *green*, and T160 in *gray*. *Blue* and *red curves* strongly fit each other. Corr gives correlation, and RMSD root means square difference between the lines corresponding to colors of symbol '×.'

horizontal. Although vertical cells become slightly wider, the magnitude of vertical mixing stays nearly stable.

5.3.4 Analyzing OWF Effect on the Ocean due to Wind Forcing Based on the Broström Approach

Section 4.2.1 introduces a simple approach to describe wind reduction behind one cubic wind farm developed by Broström (2008). That approach can be used for relative big wind farms. The here used wind farm located over four grid cells, thus comprising an area of 6 km × 6 km, almost complies with a large wind farm, compared to the used wind farm of 0.15 °E (~9.56 km at 55 °N) in the LOIZ report 2010 (Lange et al. 2010), where the Broström approach is employed. Although effect on the ocean using the Broström approach for wake description is documented, the approach is not fully evaluated so far. A comparison between wind forcing considering OWF effect by the METRAS approach and the Broström approach enables the identification of contrasts. Simulations considered for that analysis are the master simulation using METRAS forcing *T012ug08 TS01HD60F01*, here abbreviated and denoted as F01, and the run *T012ug08 TS01HD60F04*, denoted as F04. Thus ocean conditions are the same, and only forcing differs in dimension, formation, and intensification of wind wake.

The differences between run OWFr and REFr of ocean variables, surface elevation ζ, velocity components u and w, and temperature, due to the two different wind wake approaches after 1-day simulation, are documented in Fig. 5.31. It is apparent that the Broström forcing (F04) also results in the common OWF effect on the ocean. In the case of F04, the ocean shows the velocity wake, the dipole structure of surface elevation ζ, and the up- and downwelling cells connected with cooling, respectively warming. But the formation and dimension of the occurred OWF impact on the ocean system differs obviously in comparison to the METRAS forcing (F01).

In case of velocity *u-component*, the velocity wake by F04 is 0.092 m/s deeper than in F01 based on extrema at surface, but the wake is spanned nearly along the W–E cross section through the OWF (Fig. 5.31). A maximum reduction of the *u*-component by F04 is placed within the OWF at the southeasterly grid box (OWF consist of four grid boxes), but the wake trail is totally dislocated, compared to F01, which has its maximum reduction in the northeasterly grid box of the OWF district. In the case of F04, wake slightly tends to SW direction, but geostrophic effect is limited due to the short wake length downstream of the wind farm. Based on satellite data, mentioned in Chap. 4, the Broström approach underestimates the wake dimension. The stronger wake magnitude after 1 day of simulation is explainable by the compact and locally more limited effect on the ocean.

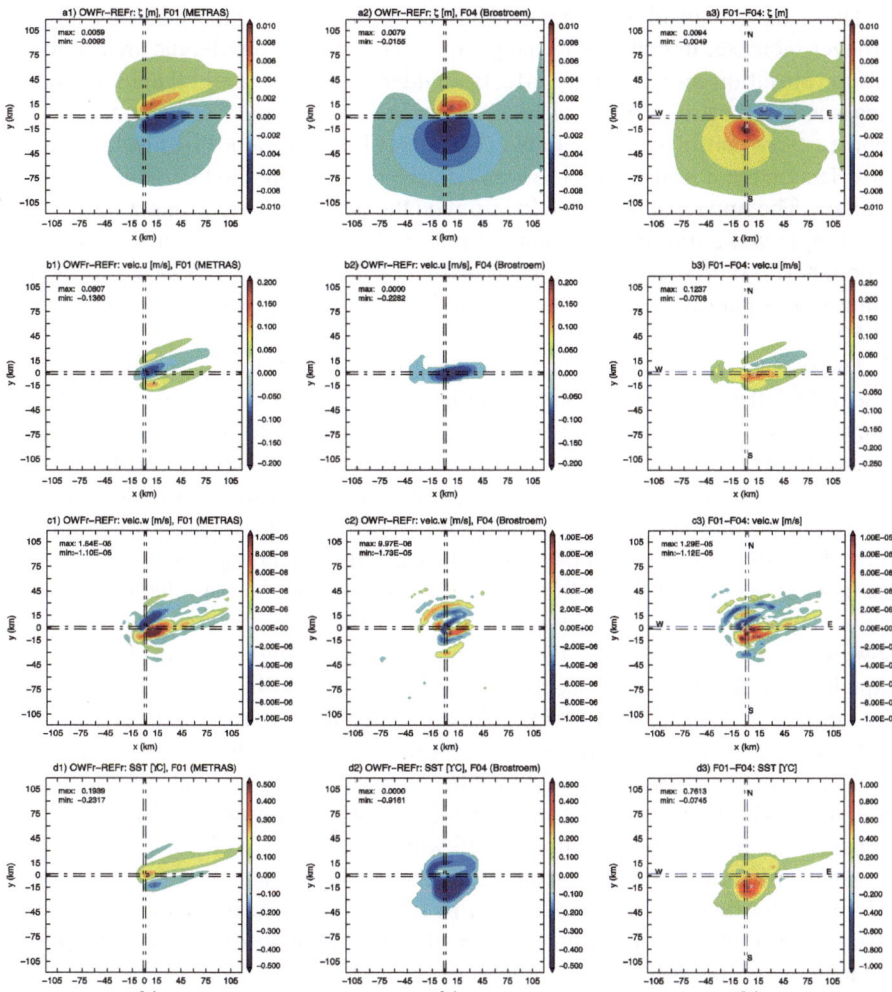

Fig. 5.31 Comparison of simulated OWF effect on the ocean using (**a1**)–(**d1**) forcing F01 (METRAS approach) and (**a2**)–(**d2**) forcing F04 (Broström approach). Illustrations (**a3**)–(**d3**) give the difference between F01 and F02. Ocean variables are (**a1**)–(**a3**) surface elevation ζ, (**b1**)–(**b3**) velocity component u at the surface, (**c1**)–(**c3**) velocity component w at 3-m depth, and (**d1**)–(**d3**) sea surface temperature SST

A more detailed view of the *surface elevation* reveals that the formation of ζ, in the case of F04, varies in dimension, compared to F01, which is connected to the dependency of the velocity wake and ζ.

Considering the whole model area, the significant positive effect on ζ is spread over a smaller area in the case of F04, even though the affected area comprises a greater uniform/denser spaced increase than what F01 shows. However, the location of ζ extrema differs by maximal two grid boxes, so 6-km distance.

The ζ maximum of F02 and F04 is biased by 2.7E-4 m, while F04 has the stronger increase. In opposite to the positive part of the ζ dipole, the area of lower surface elevation more strongly affects a wider area, thus nearly the whole model area southerly of the OWF, while changes in F01 are more concentrated at the southeasterly part of the model area (Fig. 5.31a1–a3). Again, F04 also dominates the effect on the negative ζ dipole with -0.016 m, compared to F01 with -0.009 m changes. The impact of the wind direction on the ζ formation is stronger by F01 due to a longer wake trail downstream behind the OWF. In the case of the Broström approach (F04), that effect is more or less neglected, which leads to a ζ formation being nearly parallel to the cross-section W–E, having only an inclination to it of 13.50°.

The occurrence of *vertical motion* is identified in both forcing cases, but in the case of F04, the downwelling and upwelling are not described by two main cells—a blurred transition of three downwelling zones and three to four upwelling zones around the OWF are established after 1 day of simulation (Fig. 5.31c1–c3). The induced downwelling by the OWF in the area of the positive ζ dipole results in flanked upwelling and hence, again, in downwelling zones. The downwelling is $-1.73\text{E-}05$ m/s stronger for F04 than in the case of F01, having $-1.1\text{E-}05$ m/s. But the upwelling is about 5.43E-06 m/s weaker by F04. Due to that, the difference in the velocity component w between F01 and F04 (Fig. 5.31c3), the dominant upwelling cell of F01, and the dominant downwelling areas of F04 can be explained. In the vertical along the S–N cross section, in Fig. 5.32, the impact on the vertical motion in case F04 significantly includes more intensively affected vertical layers, especially within the OWF, than F01 due to horizontal distribution of ζ, the u-velocity wake, and thus the distribution of vertical motion in the horizontal. Therefore, the extrema of the w-component are placed in lower layers, so below the thermocline for upwelling, which is for F01 above the thermocline. Maximal differences in the w-component between F01 and F04 are in order of 5E-06 m/s for downwelling and 2E-05 m/s for upwelling, with the dominant effect given by F01. Again, upwelling is more strongly influenced by changed external model assumption than downwelling.

Although upwelling is weaker in case F04, in 12-m depth, and hence in layers above, the *SST* pictured in Fig. 5.31d1–d2 shows a cooling of -0.92 °C. Here, the SST is not triggered by vertical motion because the cooling is an effect of the declination in the surface elevation.

Though the METRAS approach yields to a greater vertical motion, the effect on *temperature* is, overall, more strongly influenced by F04. Besides the cooling of SST, the use of F04 results in typical warming and cooling formations around the OWF along the S–N cross section (Fig. 5.32b1).

The horizontal dimension of the cells of temperature changing is wider (along S–N section) in the case of F04. But the location of the cell's extrema nearly fits with run F01.

The temperature extrema of cooling are located at 10-m depths, of warming at 14-m depths in both forcing cases. The discrepancies in the horizontal count more than 3 km.

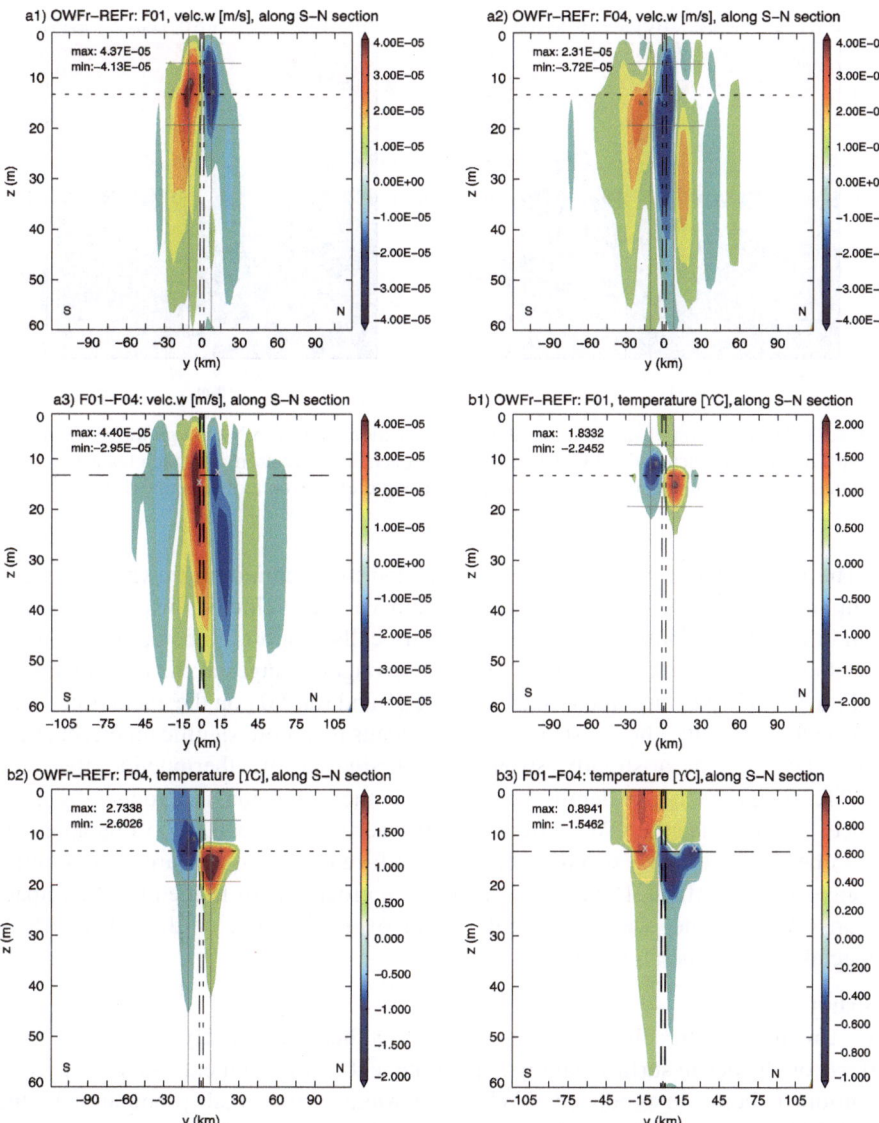

Fig. 5.32 Comparison of simulated OWF effect of (**a1**–**3**) velocity component w and (**b1**)–(**b3**) temperature along the S–N section through the OWF using (**a1**)–(**b1**) forcing F01 (METRAS approach) and (**a2**)–(**b2**) forcing F04 (Broström approach). Illustrations (**a3**)–(**b3**) give the difference between F01 and F02. Ocean variables are (**a1**)–(**a3**) surface elevation ζ, (**b1**)–(**b3**) velocity component u at the surface, (**c1**)–(**c3**) velocity component w at 3-m depth, and (**d1**)–(**d3**) sea surface temperature SST

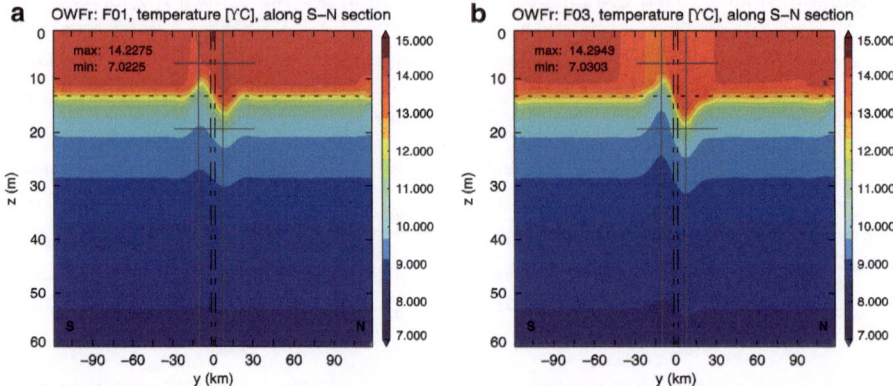

Fig. 5.33 Stratification of temperature along the y-section from S to N through the OWF for (**a**) F01-METRAS approach and (**b**) F04-Broström approach. The Broström approach leads to a more significant effect at the upper layer with slightly stronger excursion at the thermocline

The more intense warming in F04 is related to a continuous downwelling cell from surface to bottom with continuous high vertical velocities within the OWF than in case F01. Continuous vertical velocity cells support temperature advection over the entire ocean depth, which explains stronger changes at the thermocline.

Figure 5.33 illustrates the impact of run F01 and run F04 on the excursion of the thermocline. Finally, the Broström approach leads to a more significant effect at the upper layers with a slightly stronger excursion at the thermocline than the METRAS approach. At depths below 40 m, the excursion is stronger by F04; hence, F04 fortifies the OWF effect over the vertical layers. Due to this analysis, it can be assumed that a stronger effect on surface elevation forwards temperature changes in the vertical. Hence, the Broström approach helps to identify the impact of an OWF on the ocean but tends to an overestimation of the OWF effect, especially for temperature changes.

Upshot

The Broström approach (F04) was defined for theoretical analysis of wind farm effects on the ocean surface having a quadrate, cubic arrangement. Nevertheless, as mentioned, here the produced OWF wind wake is too small, compared to the METRAS approach (F01) and satellite data. However, the Broström approach is considered here to elucidate that, especially, the wind wake plays an important role due to physical reasons given in Sect. 4.2.2 and triggers the ocean system. The flanks of the wind wake, being simulated by METRAS, have a secondary role on the ocean's reaction and mainly support an upwelling zone. Anyhow, the use of the Broström approach is restricted and not as realistic as the use of the METRAS approach. As mentioned in Sect. 4.2.2, the METRAS approach has a substantial advantage where turbine specifications are considered and OWF formation is arbitrary, which ends in a more realistic wind field and therefore into a more realistic change in surface elevation. Thus, the Broström approach cannot be adapted in the case of a complex OWF arrangement. Hence, for studies of the North Sea under planned OWF construction in 2030, METRAS wind farm approach is necessary.

5.3.5 Analyzing OWF Effect on the Ocean in Case of Full Meteorological Forcing

Previous exposures of simulations dealing with the wind farm's impact on the ocean only consider effects of the forcing variables wind and pressure, indicated as the most important variables for analysis of the OWF effect on the ocean. But the whole atmosphere influences the ocean in reality, and based on the fact that wind turbines even influence temperature and humidity fields of the atmosphere, it will be investigated here how strong these additional forcing components affect the final phenomenon. But it must be noticed that the indirect forcing only allows influences of the atmosphere on the ocean and not the backward feedback. Due to the eminent change of sea surface temperature (SST), a grand question is rising and already not outstanding at all: how strong the back coupling would affect the atmosphere, especially the interaction between the ocean and the atmosphere? Based on model construction, usable infrastructure, and time scope, only the analysis of atmosphere impacts on the ocean is treated here. Full meteorological forcing for HAMSOM comprises wind speed, surface pressure, temperature, and humidity in 10-m height, cloudiness, and precipitation. Here METRAS simulation results of forcing were chosen to be cloud free without precipitation. Therefore, only temperature and humidity will have an additional impact on the ocean in that analysis. Corresponding simulations for the presentation are the master simulation *T012ug08 TS01HD60F01*, including only pressure and wind forcing, abbreviated to F01, and simulation *T012ug08 TS01HD60F03*, including full forcing, shortened to F03.

METRAS simulations are based on an SST of 15 °C and temperature at bottom of 15.59 °C. Figure 4.2 in Chap. 4 shows that a wind farm changes not only the wind but also temperature distribution, as well as humidity. As mentioned, previous results are only based on induced changes in the wind field, which mainly drives the effects in the ocean, but changed temperature and humidity conditions also impact the ocean's evaporation and heat exchange as an interaction between the ocean and the atmosphere. Such interaction won't result in a new dynamic pattern but can influence the temperature and salinity fields in the upper layers.

Heat and fresh water enters into HAMSOM through the source terms in the transport equation of temperature, respectively salinity. The source term of temperature consists of the total heat flux, which acts into the ocean at surface (into the first model layer), and from there, the effect of insolation at surface can penetrated into the ocean depths. Due to those, deeper layers can also indirectly gain heat from the atmosphere. The source term of salinity is calculated by the difference between evaporation and precipitation, while evaporation is calculated from the turbulent flux of water vapor.

In METRAS, the atmospheric boundary layer cools within the OWF due to the advection of cooler air from higher atmosphere layers to layers below the turbine rotation disc of METRAS wind turbine parameterization.

A comparison of the ocean's temperature stratification built without (F01) and with (F03) full meteorological forcing after 1 day of operating wind turbines is given in Fig. 5.34. That cooling can be finally pursued in upper ocean layers. Here, the difference between the simulation with full meteorological forcing (F03) and without (F01) along the W–E and S–N cross sections through the OWF shows that the use of the full meteorological forcing ends in a cooling down to the thermocline round $-1.30\,^{\circ}$C. That cooling results in a drop of the thermocline from 12 to 14-m depths. Due to that, the differences between F03 and F01 in Fig. 5.34c show positive values around 12 m. The nearly homogeneous upper layer of temperature in F03 (Fig. 5.34b) with a mean value of 12.5 $^{\circ}$C spreads from top to 14-m depth, while in F01, the upper layer of an average of 14.0 $^{\circ}$C ends above the thermocline, and in 12-m depth the temperature is 12 $^{\circ}$C. The warmer region within the OWF above the thermocline in the case of F03 is a result of the velocity wake. The use of F03 stamps mightiness of the upper temperature layer. That fact and the cooling involve a shift of OWF effects extrema.

Figure 5.35 illustrates the velocity components u and w, temperature, and salinity at three/two points within the OWF, 12 km southerly and 6 km northerly to the OWF along the S–N cross section over depth. Especially in the downwelling region, that switch is apparent. The minimum of velocity component w is placed at the thermocline in 12-m depth for F01, and in the case of F03, minimum is located at 14 m. The same behavior is given for temperature, Fig. 5.35c.

At position $P1$, within the upwelling region, maximal vertical velocities are placed at 11 m for F01 and at 13 m for F03, but the corresponding temperature minima are both in 10 m due to the temperature exclusion of the same intensity at this point. Simulations with F01 and F03 are correlated with 0.70 for temperature, 0.96 for w-component, 0.7–0.9 for u-component, and 0.57/0.80 for salinity. Differences at u-component occur at the wake flanks biased by 0.001 m/s, vertical velocity component is biased by 10^{-6}–10^{-7} m/s, temperature has a bias of $-0.13/0.19\,^{\circ}$C, and salinity has a bias of 0.008, especially in the upper layer. As mentioned, changes of the ocean's temperature are connected with the forcing air temperature, and due to the humidity forcing, which influences the ocean's evaporation, the salinity concentration changes, although there is no precipitation.

Based on that statistics at reference positions the effect of run F03 on the hydrodynamics is weak in comparison with the effect on the hydrographic conditions, especially for the upper layers.

Analysis of values in the horizontal at different depths, pictured in Fig. 5.36, underlines that statement. Figure 5.36 illustrates the condition of the ocean system at investigation positions through the model area chosen by considering extrema.

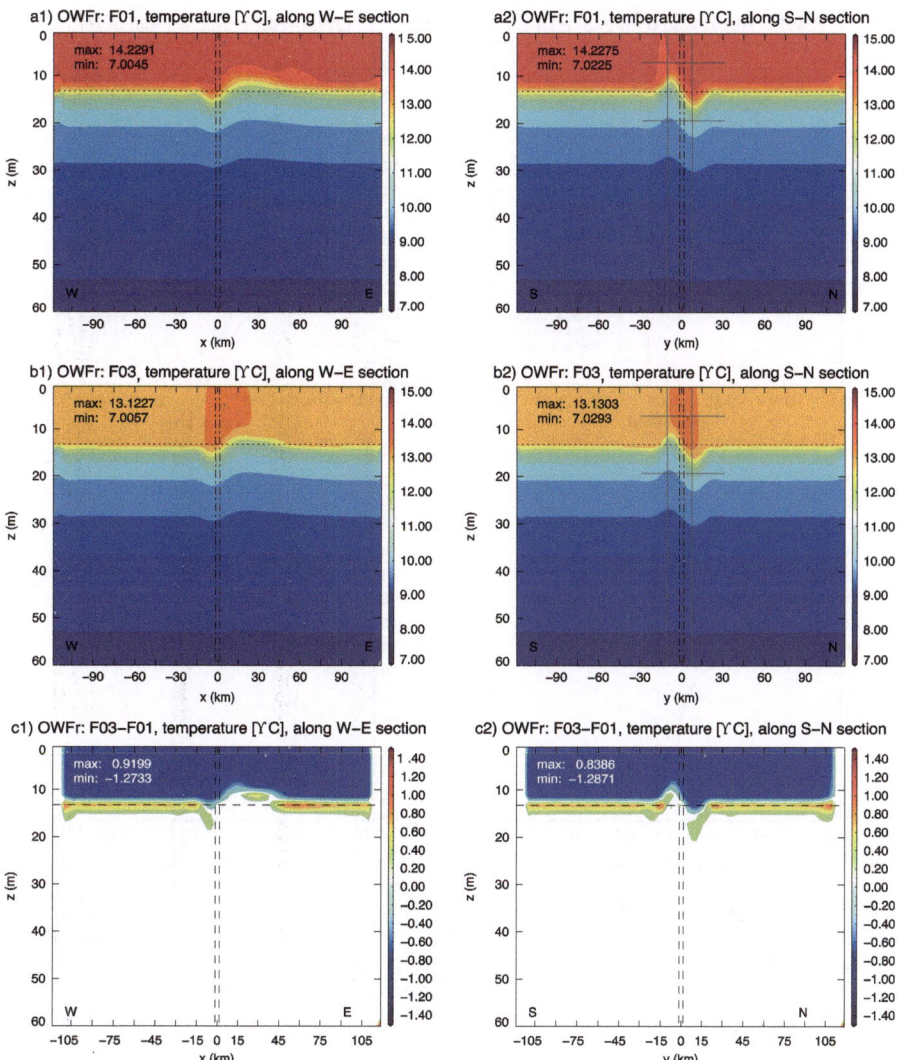

Fig. 5.34 Temperature profiles along the x-section from (**a1**)–(**c1**) W to E and along the y-section from (**a2**)–(**c2**) S to N of the restricted forcing F01 (a), the full meteorological forcing F03 (b), and the difference of both (c). Considering full forcing, the upper layers become cooler by around 1.2 °C, while close to the thermocline the layers are warmer by 0.8–0.9 °C

Surface elevation and u-component at the surface correlate run F01 and run F03 with 1.00, v-component of run F01 and run F03 is correlated by 0.94 with RMSD in order of 10^{-3} m/s.

The vertical velocity component w in 12-m depth is highly correlated by 0.99 and shows RMSD along the positions of 2.43E-06 m/s. The contrasts in the horizontal velocity components, obviously in v-component, with depth are a result of changes in the temperature, salinity field, and the density field in run F03.

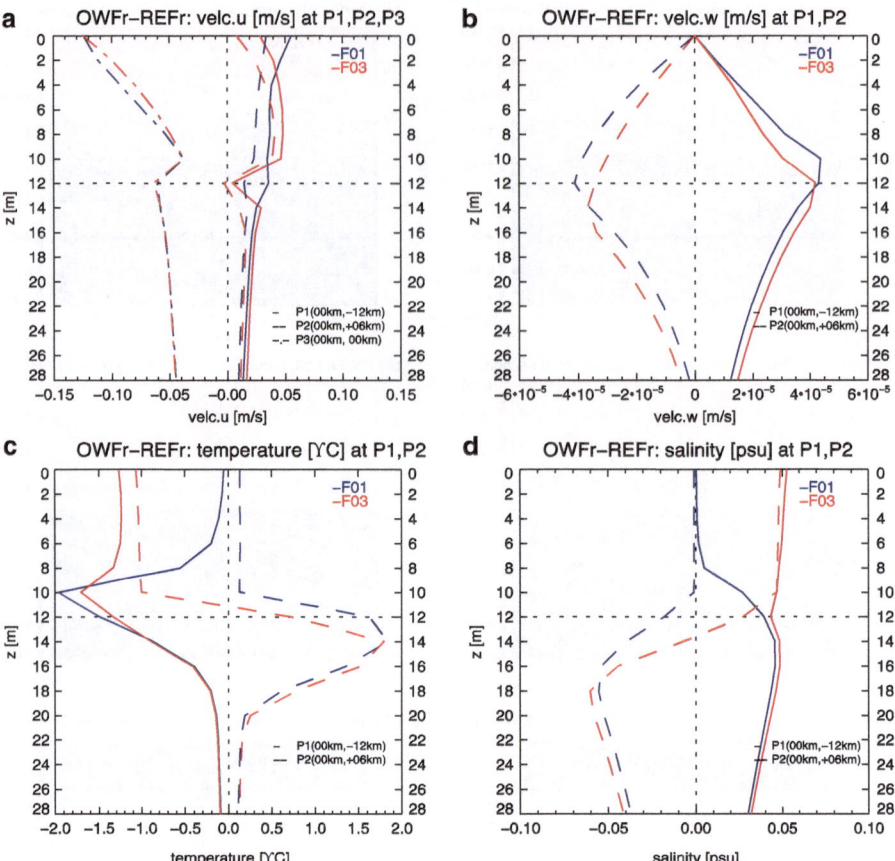

Fig. 5.35 Comparison of OWF effect on velocity components (**a**) *u* and (**b**) *w*, (**c**) temperature, and (**d**) salinity based on forcing F01 (*blue*) and F03 (*red*) over depth at two, respectively for *u*-component three, positions. *P*1(0,−12) (*solid lines*) is placed 12 km southerly of OWF and *P*2(0,6) (*dashed lines*), 6 km northerly. *P*3(0,0) (*dashed–dotted lines*) for *u*-component is positioned within the OWF. *Horizontal dotted line* marks depth of the thermocline; vertical one separates positive and negative values

The cooling of the upper layers dominates the differences in hydrographics of F01 and F03 at 12-m depth. Here, the temperature is correlated by 0.83, salinity only by 0.25 due to discrepancies within the OWF and at $P+1$ in Fig. 5.36g. Accordingly, high are the RMSD of 0.45 °C and 0.017 psu along the positions of investigation. Below 12 m, the full forcing has only a weak impact on ocean systems.

Figure 5.37 documents that ocean variables at 16 m over investigated positions are highly correlated by 0.93–1.00. While *w*-component and temperature have the best correlations, salinity gives the worst one. But from 18-m depth down to bottom, the OWF effect of run F03 on salinity adjusts to the OWF effect of run F01.

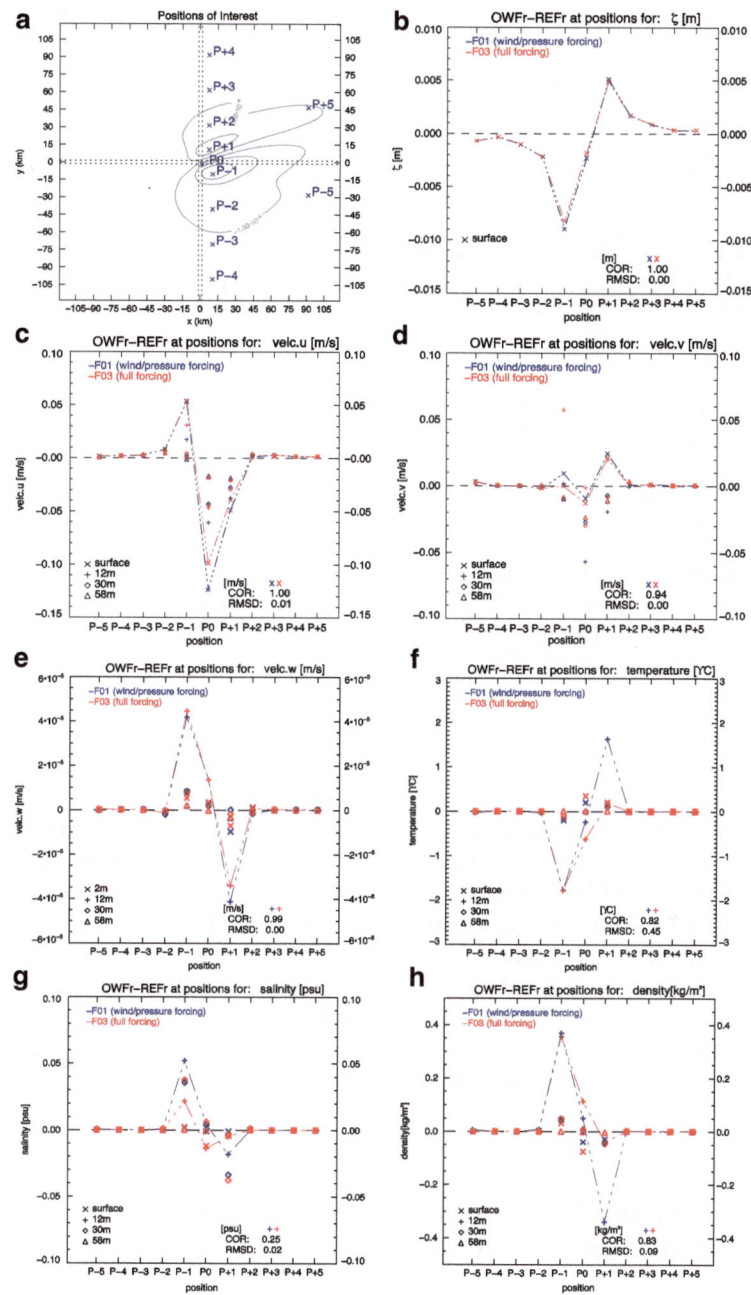

Fig. 5.36 (**a**) Location of investigation positions of interest, contour show surface elevation of simulation with F01. (**b–h**) Horizontal analysis of the OWF effect on ocean variables (**b**) surface elevation, (**c/d/e**) velocity components *u/v/w*, (**f**) temperature, (**g**) salinity, and (**d**) density) after 24 h of operating turbines at positions marked at *left top*. Results at surface (respectively at 2 m for *w*-component) '×,' 12-m depth '+,' and bottom '◊.' Correlation and RMSD are given at surface for (**b–d**) and (**e–h**) for 12-m depths. Analysis shows values for simulation with forcing F01 (*blue*) and F03 (*red*)

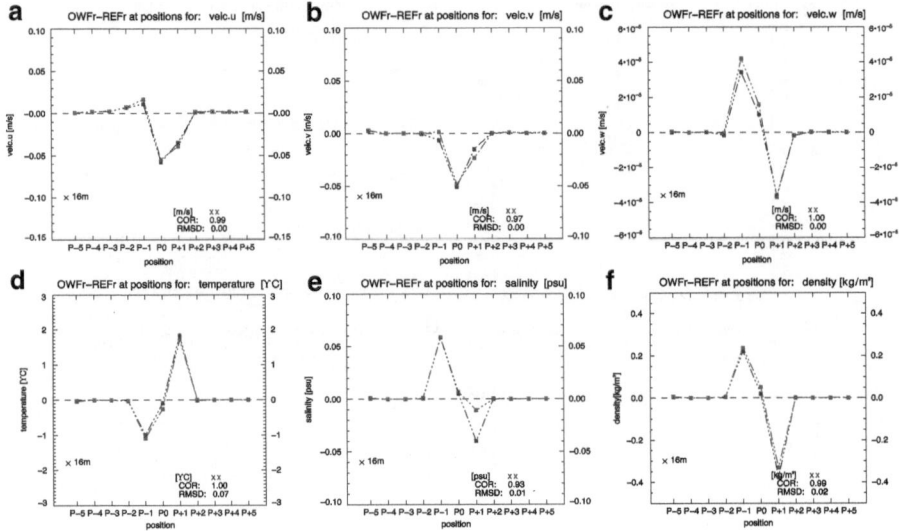

Fig. 5.37 Comparison of OWF induced changes on ocean variables based on F01 (*blue*) and F03 (*red*) over positions defined in Fig. 5.36 at 16-m depth. Ocean variables are (**a/b/c**) velocity components *u/v/w*, (**d**) temperature, (**e**) salinity, and (**f**) density. COR gives correlation, and RMSD gives the root mean square differences along positions

Upshot

As expected, meteorological forcing mainly affects the upper layer of the ocean. While the impact on dynamics is very weak, except at depth of the thermocline, the impact on temperature and, especially, on salinity is dominant. While temperature discrepancies occur till 12-m depths, the effect on salinity is stronger and includes layers till 18-m depths in the downwelling area. The use of forcing F03 finally decreases upper temperatures, which reduces gradients and weaken vertical exchange, while the drop of thermocline from 12 to 14 m increases vertical exchange in layers below. F03 has no impact on the spatial dimension of the OWF's effect on the ocean mainly because the surface elevation is equal to F01. Merely the drop of thermocline also drops extrema of vertical cells.

5.3.6   Analyzing OWF Effect on the Ocean Depending on the Depth of the Ocean

Variations in ocean depth constitute a barrier for offshore wind farm construction regarding fundaments and underwater installations. Engineering tests of swimming fundaments for wind turbines like the the Hywind Project 2009 of the Norwegian oil combined with StatoilHydro; the WindFloat Project 2011 at the coast of

Portugal; the Windflow Project 2012/2013 in France; the SWAY concept of Inocean, Shell, and Statkraft; and others will result in offshore wind farms being independent of ocean depth in the future, but such constructions are still in the testing phase. However, companies yet prefer nonswimming fundaments, and so they are restricted to shallow waters (BWE 2013). Considering a possible construction in the German EEZ, a maximum depth of 60 m, which was used in previous analysis, must be negotiated. Common depths of areas being selected for wind farming count around 30 m, like the depth of the area of OWF *alpha ventus*.

Keeping in mind that in the case of 60-m ocean depth the OWF impacts the whole ocean, this section will clarify whether a shallower water of 30-m depth will strengthen the OWF effect on hydrographic conditions and vertical mixing or not. The idea behind this analysis is that a smaller vertical extent supports a stronger vertical temperature excursion.

Analysis covers the master simulation *T012ug08 TS01HD60F01*, here denoted as HD60, and the HAMSOM run *T012ug08 TS01HD30F01*, denoted as HD30. To compare simulations based on 60-m (HD60) and 30-m (HD30) depths, the start field of temperature-salinity stratification in the case of HD30 is in accordance with the upper layers of run with 60 m because the TS start field is just cut at 30 m. So distribution from top to bottom till 30 m is the same in both cases.

A comparison of the OWF effect on surface elevation in dependence on HD60 and HD30 after 1 day of simulation is pictured in Fig. 5.38. The simulation of HD30 results in a stronger growth of the dipole effect with a difference of $+2.10 \times 10^{-3}$ and -3.94×10^{-3} m. The stronger stamped dipole in HD30 is connected with the shallower model box setup and a more intensified wake in the flow through all ocean layers, displayed at the velocity component u in Fig. 5.39a1–a3.

The reduced effect at u-component between 10 and 12 m is based on a weak reverse flow in REFr, as well as in OWFr, due to exchange processes at the thermocline. The whole ocean depth in HD30 undergoes an average of 16.14 % stronger reduction of the flow than HD60, considering only the upper 30 m for HD60. Thus, wake in the velocity component u is formed quicker, which supports an accommodation of speed between top and bottom layers and so a reinforcement of the wake.

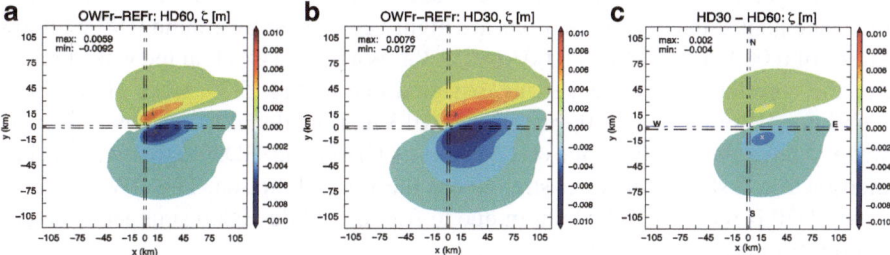

Fig. 5.38 Effect of 12-turbine OWF on surface elevation of simulation with 60 m (**a**), 30 m (**b**), and difference of 30–60 m (**c**). A stronger reaction of shallower waters is illustrated. Crosses mark the position of extrema. *Horizontal black lines* encase OWF district

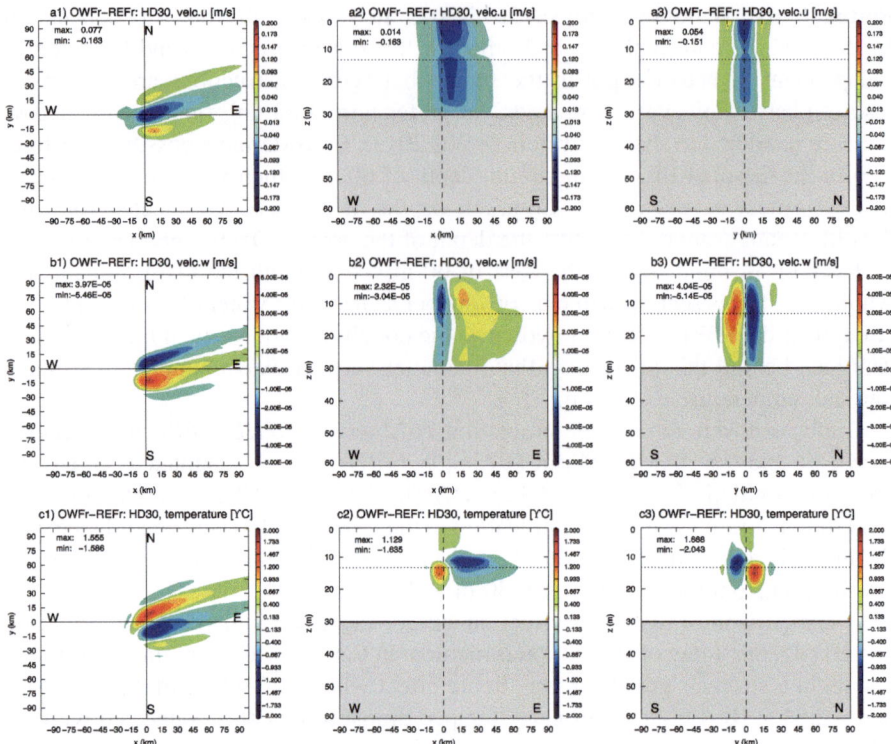

Fig. 5.39 Effect of 12-turbine wind farm on the ocean with depth of 30 m. Shown variables are velocity component u surface (**a1–a3**), velocity component w (**b1–b3**), and temperature (**c1–c3**). Illustrations (**a1**)–(**c1**) picture horizontal effect at the surface, respectively 2 m for velocity component w; (**a2**)–(**c2**) gives the x-section from W to E, and (**a3**)–(**c3**) y-section from S to N through the OWF along *solid lines* in horizontal plot. OWF is placed around $P(0,0)$

Figure 5.40a illustrates discrepancies of the u-component between simulation HD60 and HD30 at three points, within the OWF ($P3$), south ($P1$) and north ($P2$) of OWF along cross-section S–N through the OWF (exact positions of $P1$, $P2$, $P3$ are illustrated in Sect. 9.2). The points are chosen based on the position of extrema. At the wake flanks, the two simulations are highly correlated by around 0.97 with low biases of 0.009 m/s at $P1$ and 0.003 m/s at $P2$. Within the OWF in the wake area, at $P3$, the correlations are weaker, as expected due to Fig. 5.39a1–a3. Here, 0.898 correlates HD60 with HD30, biased by 0.013 m/s, while HD30 shows a stronger wake. Above thermocline (above 12 m), runs HD60 and HD30 agree well; below thermocline, a shallower water strengthens the vertical motion. The OWF effect in case HD30 is reduced at the bottom due to friction, while HD60 is unpersuaded by bottom friction in that depth.

The occurrence of vertical motion has the same distribution in the horizontal for both ocean depth cases. Also, the vertical cells range from surface to bottom in both cases, but in run HD30, the vertical cells are smoother in the vertical and clearly

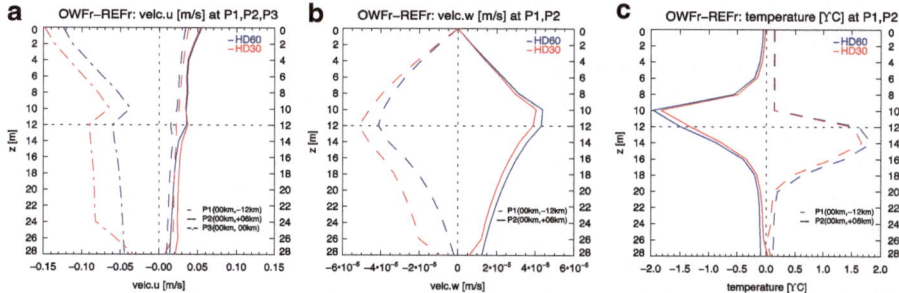

Fig. 5.40 Comparison over depths of velocity components (**a**) u and (**b**) w and (**c**) temperature for simulation based on 60-m (*blue*) and 30-m (*red*) depths at positions in km $P1(0,-12)$, $P2(0,6)$ and for velocity component u, additional $P3(0,0)$. *Horizontal dotted line* marks the depth of the thermocline; the vertical one separates positive and negative values

result in two dominant up- and downwelling cells around the OWF. Especially along W–E section through the OWF (Fig. 5.39b2), the upwelling is undisturbed with depth and more consistent in HD30 than in HD60. In case HD60 occurs after 1 day of simulation a zone of downwelling below the thermocline and eastward of the OWF. In case HD30, the effect shown at 12-m depth (Fig. 5.39a) can be broaden from surface to bottom. Simulation HD30 has got a more intensive downwelling of maximal -5.46×10^{-5} m/s, compared to HD60 with -4.37×10^{-5} m/s over the whole model box, and also the mean of downwelling over the affected area leads to a stronger effect by HD30 with -1.27×10^{-6} m/s than HD60 with -1.09×10^{-6} m/s. The depth where the extrema occur is 12 m in both cases. The maximal upwelling occurs in depth of 10 m for both cases, and again the run HD30 shows here a stronger value being 0.11×10^{-6} m/s higher than in case HD60. Below 10 m, the upwelling is weaker than in 10-m depth for both simulations, but the run HD60 becomes here more intensified due to bottom friction affecting results in run HD30. In the horizontal, HD30 tends to location of extrema through layers being closer to OWF grid boxes with maximal differences to HD60 by 6 km in x-direction, so two grid boxes, and 3 km in y-direction; location of extrema of H60 is slightly more easterly positioned after 1-day simulation. Inspection of distribution of w-component at points $P1$ and $P2$ (Fig. 5.40b) shows that downwelling at $P2$ is weaker in HD60 while upwelling at $P1$ is nearly identical till 4-m depth. Below 4 m and, especially, from 10 m on, HD30 gives a lower upwelling.

One must note that the position of investigation $P2$ does not fit with the position of overall maximal upwelling; that is why run HD60 dominates here in the upper layers. The dominant downwelling for HD30 is registered for the whole 30-m ocean depth. Correlations of w-component of runs HD60 and HD30 are 0.98 in $P1$ and 0.95 in $P2$, while again main discrepancies occur below the thermocline due to different defined depths of the model bottom. The two simulations are more highly biased for downwelling by 9.15×10^{-6} m/s, while upwelling is only biased by 4.5×10^{-7} m/s.

In dependence of vertical motion, two zones of changes in the temperature field are obvious (Fig. 5.39c1–c3). Although the vertical velocity component w in run HD30 is greater for downwelling and mostly for upwelling, the maximal effect of the temperature within the model box is more dominant in case HD60, but discrepancies in the global mean change between run HD30 and over the upper 30 m of HD60 counts only $-0.0024/+0.004$ °C for cooling/warming. The vertical position of extrema occurs in both cases around the thermocline in 10-m depths for cooling and in 14-m depths for warming. A global maximal change in temperature is $-2.70/+1.92$ °C for run HD60 and $-2.36/+1.68$ °C for run HD30. In the case of HD30, the effect on temperature in the vertical is more strongly located around the thermocline due to a decrease of w-component with depths below 12 m. Due to smoother vertical cells in HD30, the change of temperature is more uniform over the affected areas than in case HD60, which leads to a smaller global mean over the whole ocean depth for HD60, compared to HD30. However, a change in the temperature over the vertical at positions $P1$ and $P2$ (Fig. 5.40c) shows the little more dominant effect on the temperature field by HD60. Temperatures at both positions are more strongly correlated, with 0.99, than velocities' w-component. In the upwelling region occurs the highest bias of 0.060 °C, compared to 0.004 °C for the downwelling position, due to the fact that in case HD60, advection of cooler water, below 30-m depths, cools the upper layers.

Upshot

Summarizing, shallower water depths strengthen the wake in the u-velocity, and hence a stronger dipole structure of surface elevation occurs. Therefore, we can expect a stronger downwelling and also a stronger upwelling above the thermocline in shallow waters. Here, the vertical positions of w-component extrema are independent of the ocean depth but strongly depend on ocean stratification. Also, a shallower water leads to stronger distinct vertical cells from top to bottom, while in the case of deeper ocean, the formation of the vertical cells vary more in the horizontal. Nevertheless, the ocean depth plays a secondary factor for the common OWF impact on the ocean system. Like in the previous analysis, it becomes clear that the distribution of hydrographic conditions and position of thermocline are more significant for the OWF effect because at the thermocline the OWF induced dynamical change effectively impacts the ocean system.

5.4   Evaluation of Modeled OWF Effect on the Ocean

The theoretical approach of using HAMSOM over an ocean box to determine the effect of an OWF on the ocean's dynamic gives possible dynamical changes. Although HAMSOM is a well physically proofed model, the here used restrictions of a model box and its forcing lead to the question on how realistic the dimension of arising phenomenon, treated in the last sections, is. Hence, a snapshot of conditions around the offshore wind farm *alpha ventus* was taken owing to BSH's support,

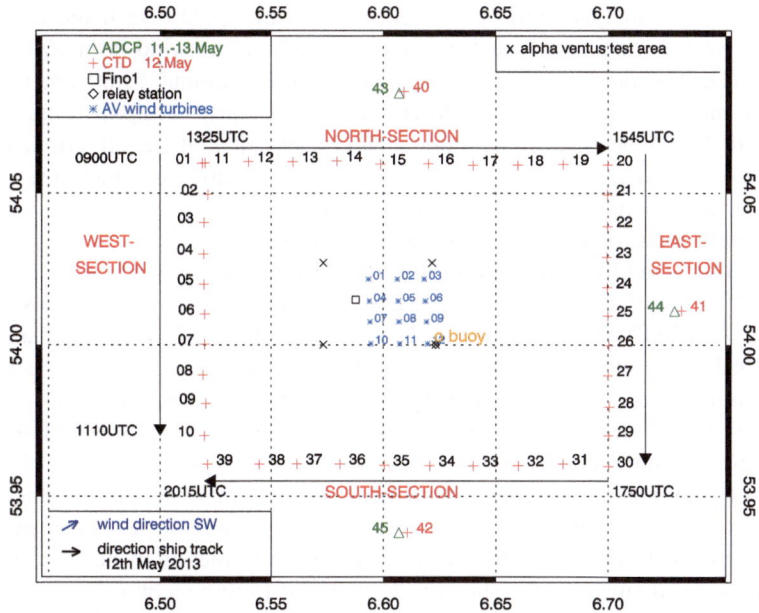

Fig. 5.41 Map of investigation area around test wind farm *alpha ventus* (*black* '+'). Wind turbines are *marked blue*, CTD measurements are *red* marked with '+', ADCP measurements *green* marked with 'Δ', the sign for the Fino1 platform is '□', and for the relay station '♦', which is close to turbine 12 and the swell buoy (*orange* 'o buoy'). *Black arrows* mark direction of ship track for CTD measurements. The CTD sections (west, north, east, south) are defined along the CTD stations

which supports to put model results and measurements of temperature and velocity into relation to each other.

The area around wind farm *alpha ventus* with measurement stations, time, and position of data collection are represented in Fig. 5.41. The measurements comprise 39 + 3 CTD stations and three ADCP mooring stations only taken for this analysis. The time frame was May 11–13, 2013, at which CTD measurements were taken on May 12 and the ADCP instruments collected data over 2 days from May 11–13. Additional temperature and dynamical data were and are retrieved permanently by instruments of the station Fino1, relay station, and swell buoy located in the area of *alpha ventus* (Fig. 5.41).

A separation of ADCP-measured velocity data into its component was deemed as the easiest way to detect upwelling and downwelling, as shown in the model data. But the residual velocity signal in the North Sea is strongly disturbed by the tide, which makes the analysis of velocity components difficult, as well as the fact that changes of the vertical velocity component are small and hard to detect, even after subtracting the tidal signals. Hence, the ADCP data taken at three positions (north, east, and south to *alpha ventus*) do not result in a distinct, with model data comparable, signal, and thus they are only documented in Sect. 9.1.3, for the sake of completeness.

Onward, the analysis will be concentrated on the CTD data taken along four sections westerly, northerly, easterly, and southerly of the wind farm *alpha ventus* (Fig. 5.41). The northern and southern sections are around 12.80 km long, the western and eastern ones around 10.90 km. The distance to *alpha ventus'* center counts in longitude direction around 5.5 km and in latitude direction, 6.4 km because of bordering prohibited zones based on wind farm constructions.

The arrangement of investigated locations was chosen based on previous model results, with focus on catching modeled OWF effects.

The used model simulation for the evaluation is the run *T012ug08 TS03HD30F01* of the model setup TOS-01 (ocean box), which means a wind farm of 12 wind turbines (T012) within 4 grid cells, a wind forcing based on a prescribed geostrophic wind of 8 m/s (ug08), a temperature and salinity start field based on the measured profiles during the ship cruise (denoted as TS03), a model depth of 30 m (HD30), and only a forcing being established by METRAS 10-m wind field and surface pressure (denoted as F01).

Hence, the main differences have been simulation assumptions, and nature conditions existing during the measuring campaign are the use of a simplified meteorological forcing, not the exact same wind turbines (hub height difference of 10 m, rotor diameter difference of 36 m, and different technical parameters), a homogeneous water depth of 30 m, an averaged initial conditions for temperature, and salinity based on taken CTD measurements, and in the model simulations the tide is neglected.

Considering only *meteorological* pressure and wind, forcing is justified due to the fact that there was no possibility to get an area-wide realistic meteorological forcing (for example, satellite data) and due to the fact that dynamical oceanic changes are dominantly driven by wind (result of Sect. 5.3.5). The decision of using wind forcing based on a prescribed geostrophic wind of 8 m/s is leaned on to be the closest description of wind situation discovered on-site.

The wind turbine parameters were not changed for the comparison simulation due to computation time and costs, as well as because the used model resolution tends to overestimate the wind wake dimension orthogonal to the wind direction. Additionally, it must be said that during the measuring period, not even all of the 12 turbines were running all the time due to planned maintenance, and regarding *alpha ventus'* power plan, the turbines were not running with full power. Based on these facts, it is supposed that the used smaller turbine may balance the later listed issue of horizontal resolution and that the used wind turbine adjustment fits reasonably with reality.

The *bathymetry of model* is flat, which is not a big limitation since the investigation area around *alpha ventus* is known as flat and sandy with an average depth of 30 m in marine charts. Such a bottom topography was chosen because the dependence of OWF effect on the water depth is relatively low, as shown in Sect. 5.3.6.

It must be clear that this section provides an evidence of physical accuracy of model results and gives an estimation of OWF effect's dimension by means of temperature analysis.

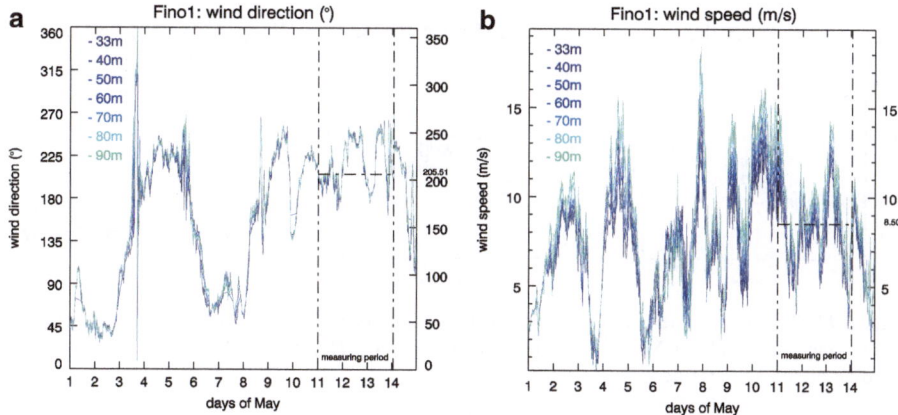

Fig. 5.42 Wind situation in May 2013 at Fino1, including measuring period from 11 to 13 May. *Left* shows wind direction; *right* shows wind speed. During measuring period, wind direction was mainly SW with a mean wind speed of 8.10 m/s above 33 m. *Horizontal dashed–dotted line* gives median of shown variable

Prior to doing a description of the evaluation between measurements and model results, the situation in May before and during data collection around *alpha ventus* is presented using data of research platform Fino1. The important data here are information about the wind situation, which is depicted in Fig. 5.42.

The wind often veers between south and west over the days before the measuring period. Especially from May 8, 2013 on, wind directions play between 175° and 250° fairly constant over heights between 33 and 90 m. An average wind direction of 205.51° predominates during the measuring period, which nicely accords with wind direction of used METRAS 10 m wind forcing field. Even the wind speed, measured at Fino1, is 8.50 m/s, averaged over heights and campaign time. Considering METRAS 10 m wind field of 6.5–7.0 m/s (based on prescribed $u_g = 8$ m/s), we can say that the wind forcing for HAMSOM is close to the realistic wind situation. Discrepancies are kept in mind for the evaluation.

Oceanic conditions of the North Sea on May 2013 show a continuous increase of temperature as expected in springtime based on solar radiation, Fig. 5.43. During the measuring period, there occur values with averages over depth (3–25 m for temperature and 6–25 m for salinity) and time of 7.42 °C for temperature and 32.77 psu for salinity (Fig. 5.43). The mean flow velocity over depth and time was 0.44 m/s and at 2-m depth around 0.85 m/s.

As mentioned, the strong tides in the North Sea hamper the measurements of the vertical velocity component w via the used ADCP instruments; therefore, the CTD measurements play a key role in the presentation of evaluation. An advantage of the CTD data is that the temperature is not strongly affected by the tide and is expected to represent the vertical stratification with evidence of the vertical mixing, vertical exchange, and the vertical motion, as defined in the model results. Variations of the CTD measurements, especially at the surface, based on alternating cloudy

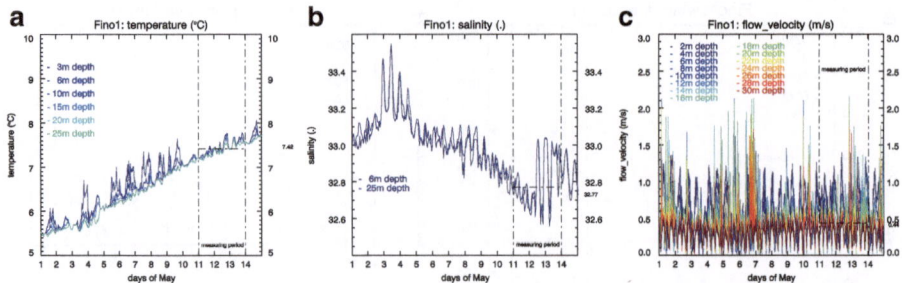

Fig. 5.43 Temperature (**a**), salinity (**b**), and flow velocity (**c**) of May 2013 at Fino1, including measuring period 11–13 May at various available depths. *Horizontal dashed–dotted line* gives median of shown variable

conditions (no precipitation) and slightly varying swell during the measuring period, are considered in the following analysis of temperature.

Supposing the offshore wind farm *alpha ventus* has no effect on ocean dynamics, we would expect standard temperature stratification with higher temperature in the upper layers, a more or less clear thermocline (no excursion), and cooler layers below. Also, and it's important, we would expect that such a stratification occurs over the whole area and is quite constant. The investigation area is placed in shallow water strongly affected by wind. Therefore, the sea around the wind farm *alpha ventus* is expected as well as mixed with mostly no or very weak thermocline, although solar radiation would support a formation of thermocline.

The CTD measurements result in a maximum SST of 8.0 and 7.0 °C at bottom on May 12, 2013. Though the weak temperature difference of 1.0 °C is from top to bottom, the CTD measurements show complex structures, which are illustrated in Fig. 5.44.

Precise structures in the vertical occur along the four sections of CTD in the west, north, south, and east of *alpha ventus*. Zones of cooler and warmer temperatures were observed along the tracks, which look like an undulating formation clarified by black dashed lines in Fig. 5.44, which is evidence of the excursion of the thermocline. All sections have in common that cooler temperature ridges border on a trough of warmer temperature values. Such a zone of slightly raised temperatures is mostly placed along the sections at the sector of *alpha ventus* between longitude and latitude positions 54°N 6.57°E, 54.03°N 6.57°E, 54.03°N 6.62°E, and 54°N 6.62°E (district of *alpha ventus* test wind farm area *x* in Fig. 5.41).

These zones, 'bubbles' from bottom to surface or surface to bottom, were measured in all sections with various intensities. The western and northern sections comprise cooler temperatures than the eastern and southern ones, with differences of tenths of degree.

Along the western section from south to north, a cooler ridge was formed with values of 7.28 °C blurring at 9-m depths (Fig. 5.44). Another cooler ridge occurred at the most northern part of this section with 7.25 °C. A slightly warmer temperature column disconnects these ridges beginning at the northern projected corner of *alpha ventus*. *At the northern section*, warmer temperatures transported to depth

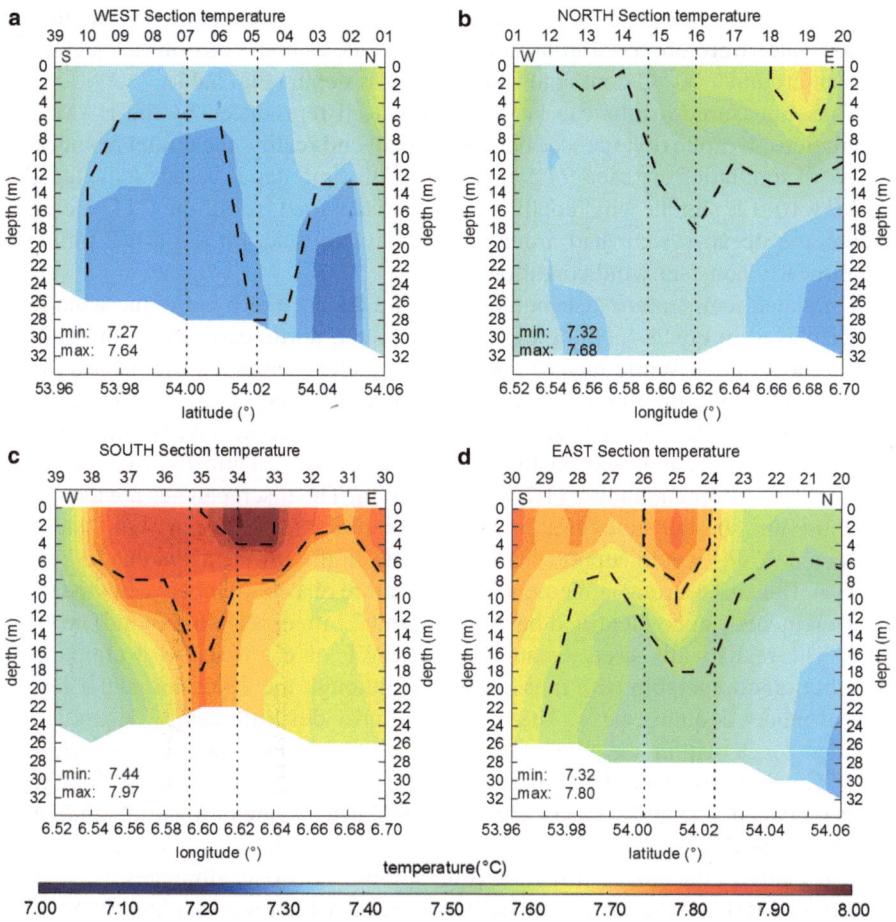

Fig. 5.44 CTD temperature sections (**a–d**) around the wind farm alpha ventus from May 12. Illustration (**a**) gives west section, (**b**) north section, (**c**) south section, and (**d**) east section. *Black dotted vertical lines* mark abreast with the wind farm. *Black dashed lines* accent temperature formation comparable with HAMSOM model results. Temperature structure shows evidence of vertical mixing due to the wind farm's wind wake. Superior *x*-axis gives the number of CTD position. Distance of sections to the wind farm counts 5.5–6.5 km. Length of latitude sections is 12.8 km and of longitude section, 10.9 km. In the horizontal, CTD was allocated without any interpolation. In the vertical, data were averaged over 2 m, which is consistent with used vertical resolution of HAMSOM

could be seen in the latitude of *alpha ventus*. A maximum was observed at the eastern part of this section, which is a part being defined to be in the wind wake of the wind farm (Fig. 5.44). Moving from north to south *along the eastern section*, temperatures increased with a maximum drop in the zone behind the wind farm *alpha ventus* of 7.76 °C. *Along the southern section*, the highest temperature values occurred with a maximum of 7.97 °C close to surface in the more eastern range of the wind farm. Temperature values of 7.8 °C reached the bottom forming a funnel at latitude of *alpha ventus*. West and east of the wind farm, local extreme 'bubbles' of cooler water were observed influencing the surface at both ends of that section. A zone of around 7.90 °C dominates down to 10-m depth (Fig. 5.44).

These measured results can be reasonably well reproduced by the HAMSOM simulations. Despite divergences between model and reality, the model results after 3 days of simulation are used for comparison. This time step is based on the fact that on May 10–12, similar wind conditions were discovered, so till the CTD measurements, the ocean system had around 2.5–3 days to react on the OWF influence under nearly constant wind conditions.

Simulated temperature distributions within a distance of 6 km to the wind farm are displayed in Fig. 5.45 along each section (west, north, east, south) through the whole model area. While measurements show a clear temperature transition from surface to bottom, the model results are blocked by a too strong thermocline as a separation frontier, which leads to nearly two main ocean layers, although the starting TS profile is based on the CTD data. A layer of mostly 7.65 °C exists above 12 m, a dominant layer of 7 °C below 20 m. The discrepancies are a result of tidal mixing. In the here used model simulation of the ocean, box tides are neglected. Tides would support vertical mixing in the lower layers due to bottom friction. But still a strong agreement of temperature distribution between model and measurements can be identified between 10 and 22 m, especially around the OWF area. The realistically used distance to the OWF of 6 km shows a temperature distribution comparable with measurements, although the 3-km horizontal resolution of model is quite coarse. The model also provides little details depending on warmer zones within the wind farm sector. These zones have temperatures of 7.73 °C, which means a difference of 0.1 °C, compared to the not OWF-affected areas (between −45 and −90 km and 45 and 90 km), and are of the same dimension as that of the CTD.

Especially in the *northern section* (Fig. 5.45b), the model simulates the warmer extrema easterly displaced from wind farm sector, which was also measured. That extremum can be identified in the model by a 6-km distance to OWF, but at 18-km distance, the structure is closer to measure (Fig. 5.44b, north section).

In the case of the *south section* (Fig 5.45c), the maximum with depth is shifted to the front of the wind farm zone, while CTD data show a maximum within the wind farm sector. The overall maximum occurred close to surface, east of the OWF zone, which can also be found in simulations. The wave formation in sections with peaks and troughs is overestimated along the sections by the model, while the structure gives a comprehensive agreement.

The *west section* (Fig. 5.45a) shows a wave formation with one minimum and two peaks, whereby the maximum peak in the south is wider and stronger then the northern one. The model results in a horizontal width of the southern peak of 15–20 km; the horizontal dimension by measurements leads to around 5 km.

At the *north section* (Fig. 5.45b), one trough is easily observable, followed by a rudiment of a long persistent peak, between 20 and 75 km, where colder water is upwelling. Over that peak, at and close to the surface, the mentioned warmer extreme is located. Here, the model gives a horizontal dimension of the trough

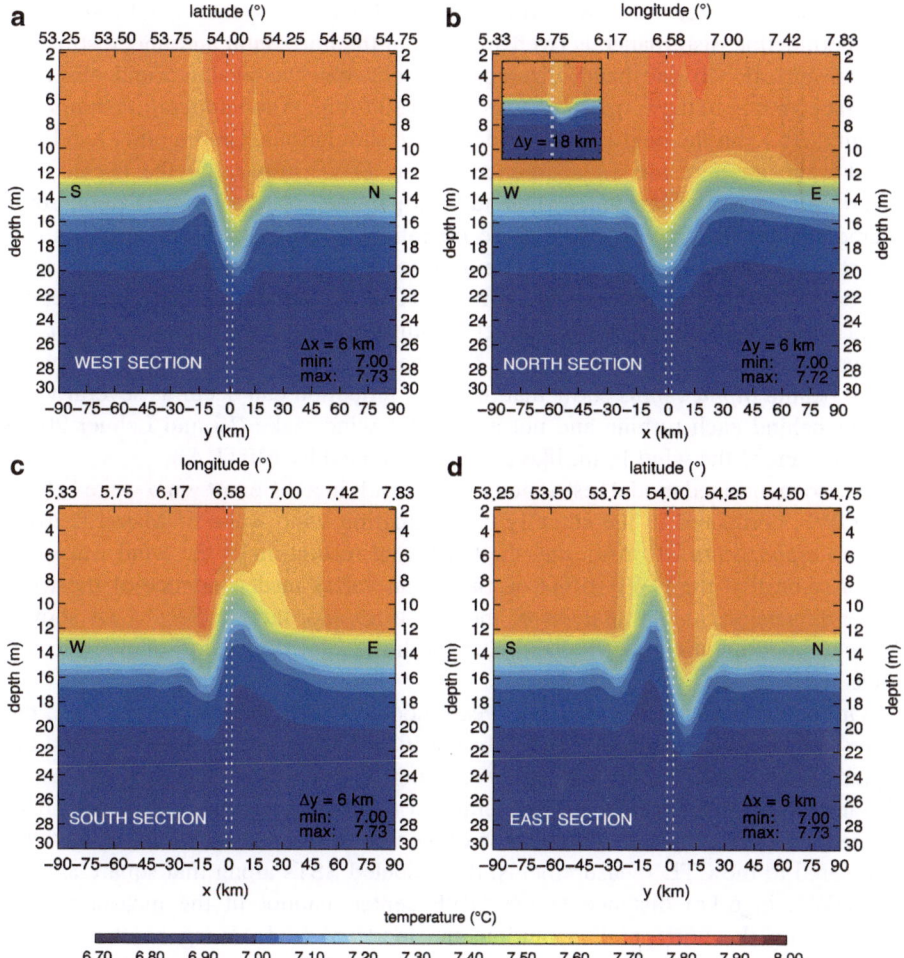

Fig. 5.45 HAMSOM's result for temperature distribution against depth along (**a**) west, (**b**) north, (**c**) south, and (**d**) east sections around the 12-turbine wind farm. Distance Δ*x* and Δ*y* of each section to wind farm is 6 km (*big figures*). *Small figure* in the north section (**b**) shows result for Δ*y* = 18 km. Result is after 3 days of simulation T012ug08TS02HD30F01. Projection of wind farm position is marked with *white dotted lines*

with 30 km, measurements only with 3 km. The local field of extrema at surface has a horizontal dimension of 7–10 km in the model, around 2 km in case of CTD measurements.

The warmer region, at the *south section* (Fig. 5.45c), spans a horizontal length of around 14 km, the model 15–45 km. The maximum within the wind farm corridor at the *east section* has similar dimensions in the horizontal for both data sets. But the southern peak has a horizontal resolution of 4–5 km for CTD data, 20 km for the model results.

The model depicts the OWF effect quite well, but due to the model restriction, the thermocline disturbance is overestimated in the horizontal and underestimated in the vertical. On one hand, discrepancies between the modeled and observed ocean occur due to the simplified meteorological forcing (no temperature, humidity forcing, etc.) and the negligence of tides in the simulations. On the other hand, the wind wake itself results in differences in the OWF impact on the ocean. The discrepancies in the technical wind turbine parameters (thrust coefficient, rotor diameter, hub height, etc.) and in the operation mode of the wind turbines can affect the wind wake description. The used forcing of a constant 10-m wind field prescribed by a geostrophic wind of 8 m/s and constant wind direction can affect the wind wake. And the horizontal model resolution of 3 km × 3 km can affect the dimension of the wind wake as well. For example, satellite analyses of the wind wake behind *alpha ventus* show that the wind reduction can occur as several wind wakes behind each turbine and not as one big wind wake (Li and Lehner 2012) downstream of the wind farm, like the one simulated by METRAS.

These mentioned model restrictions could result in a different wind wake behind the OWF, compared to the actually predominating wind wake of May 12, 2013, behind *alpha ventus*. Considering the horizontal resolution of the wind wake, we can assume that the OWF effect on the ocean varies in dependence of the wind wake dimension, as the analysis of the Broström approach in Sect. 5.3.4 shows. Hence, the simulation can overestimate the horizontal dimension of the affected areas in the ocean.

On the basis of a comparison between modeled SSTs and measured SSTs, the issue of the horizontal resolution can be clarified. In the case of measurements, the square around the OWF along the west, north, east, and south sections has a distance to the OWF center of an average of 6 km. In assumption that the model overestimates the horizontal dimension of the OWF effect on the temperature field, compared to the CTD measurements, the modeled SSTs along that square around the OWF, in 6-km distance to the OWF center, cannot fit the measurements. Considering the horizontal resolution, the modeled SSTs along another square around the OWF having a greater distance to the OWF center is provided.

Figure 5.46 exemplifies the situation of measured SSTs around the wind farm *alpha ventus* along the mentioned 6-m-distance square on May 12, 2013. The line plots summarize that the highest SST values were measured along the southern and eastern sections, and the lowest values were detected at the west section. The cube illustration helps to get a spatial idea of the SST along the square around the wind farm.

The counterpart, the modeled SSTs around the OWF, is pictured in Fig. 5.47. Figure 5.47a shows the modeled sea surface temperature of the whole model area. In the middle, the OWF district (four grid boxes) is marked with a solid black square. The modeled SST along the corresponding square around the OWF with a distance of 6 km to the OWF center illustrates that the modeled SSTs strongly differ from the measured one.

Using a greater square provides a better agreement; moving the original '6 km-distance square' more to the north and expanding it to the west and slightly to the

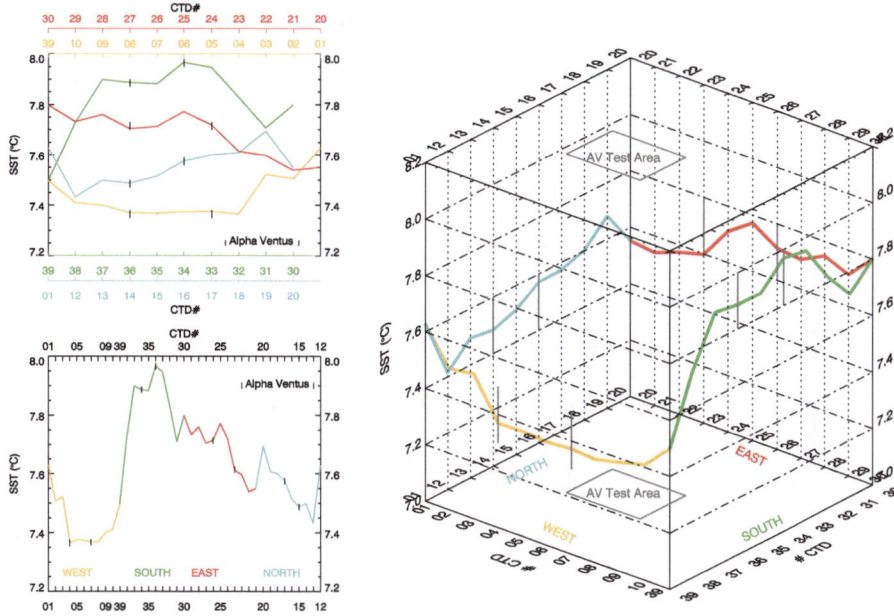

Fig. 5.46 CTD-SST around the wind farm *alpha ventus* in three different presentations to exemplify surface temperature field along sections 'West' (*yellow*), 'South' (*green*), 'East' (*red*), and 'North' (*blue*). *Gray bars* mark the section of the wind farm's position. The *upper left figure* illustrates SST for each section. The *lower left figure* accents temperature differences around the wind farm. The *right figure* helps to get a spatial idea of SST distribution. The cube's *z*-axis gives temperature in °C, starting with 7 °C and ending with 8.2 °C. *X*-axis and *y*-axis reflect CTD numbers, respectively position

east and then the modeled SSTs along that square (in Fig. 5.47c) fits quite well the measured SSTs in Fig. 5.46. Along that new square, the modeled SSTs show a drop in SST along the west section, common lower SSTs in the north, higher values in the section of the OWF at the east section, and the maximum of increased SSTs along the south section. The new square (Fig. 5.47c) counts in the north and east a distance to the OWF center of 12 km, in the south only 3 km, in the west 15 km and comprises a 2.8 times greater area than the '6 km-distance square,' which means in *x*-direction the dimension of the simulated temperature effect is overestimated by factor 1.75, in *y*-direction by factor 1.25, compared to observations. In average, the differences between observed SST along '6 km-distance square' and simulated SST along the greater square count 0.20 °C.

The modification of the square around the OWF for comparison is not performed arbitrarily but is based on known computational restrictions—here, the horizontal resolution of the wake presentation. Apart from that, the model captures the observed temperature structures.

At the corners of that square, a cooling is observed with the exception of the southeast corner. The fact that the corners show a cooling result in the assumption

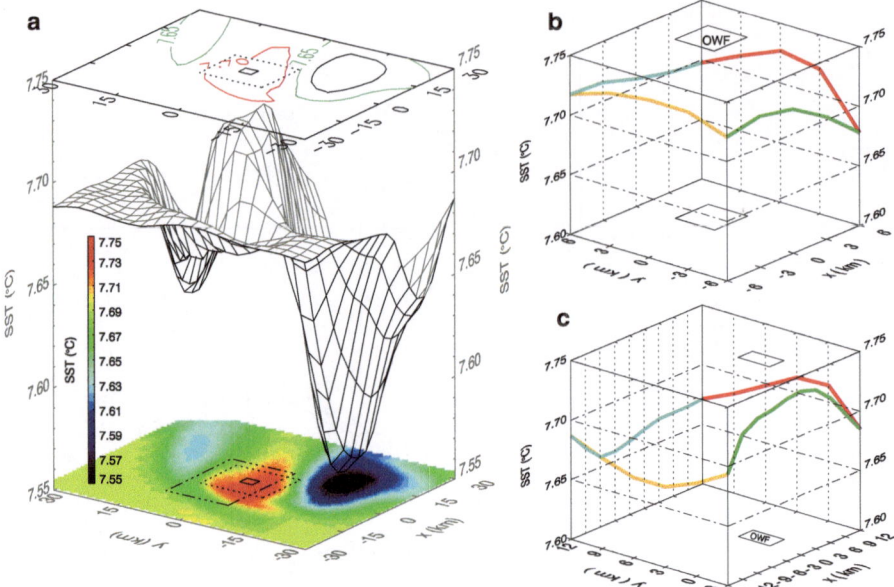

Fig. 5.47 (**a**) HAMSOM SST in three-dimensional illustration. *Square of solid lines* mark OWF in the middle of the area, which is surrounded by *dotted square*, indicating sections by 6-km difference to OWF. Additionally, a square of *dashed–dotted lines* highlights closest SST, compared to CTD measurements. Illustration (**b**) and (**c**) picture HAMSOM SST along squares of (**a**). Illustration (**b**) shows HAMSOM SST along the section in 6-km distance to OWF. Illustration (**c**) shows HAMSOM SST along the section with a distance of $\Delta y = 12$ km from the OWF to the northern section, $\Delta y = 3$ km to the southern section, and $\Delta x = 15$ km to western section, and $\Delta x = 12$ km to eastern section. *Little black squares* mark the OWF position in (**b**) and (**c**). Results are based on three days of operating wind turbine simulation

that these corners are outside of the main downwelling cell, where the model shows weak positive vertical velocities.

With using the '6 km-distance square,' it appears to be too small to detect the upwelling cell being expected south/southeast of the wind farm, as model results show.

Overall, the agreement between simulation and measurements is impressive, considering the theoretical model setup and computational restriction and the observed temperature profiles linked to downwelling within and around *alpha ventus*.

References

Backhaus J (1985) A three-dimensional model for the simulation of shelf sea dynamics. Ocean Dyn 38:165–187. doi:10.1007/BF02328975

Broström G (2008) On the influence of large wind farms on the upper ocean circulation. J Marine Syst 74:585–591

BWE (2013) Konstruktiver Aufbau von Windanlagen. Bundesverband WindEnergie

Hein B (2013) Processes of stratification and destratification in the Mekong ROFI. Dissertation, Universität Hamburg 156

Jones D (2010) Diffusion MRI: theory, methods, and applications. Oxford University Press, Oxford

Lange M, Burkhard B, Garthe S, Gee K (2010) Analyzing coastal and marine changes: offshore wind farming as a case study. LOICZ Res Stud 36:212

Lax P, Wendroff B (1960) Systems of conservation laws. Commun Pure Appl Math 13:217–237. doi:10.1002/cpa.3160130205

Li X, Lehner S (2012) Sea surface wind field retrieval from TerraSAR-X and its applications to coastal areas. In: IEEE international geoscience and remote sensing symposium (IGARSS), IEEE, pp 2059–2062

Moum J, Smyth W (2001) Upper ocean mixing processes. Encyclopedia of Ocean Sciences 3093–3100. doi:10.1006/rwos.2001.0156

Paskyabi M, Fer I (2012) Upper ocean response to large wind farm effect in the presence of surface gravity waves. Energy Procedia 24:245–254. doi:10.1016/j.egypro.2012.06.106

Pohlmann T (2006) A meso-scale model of the central and southern North Sea: consequences of an improved resolution. Cont Shelf Res 26:2367

Roe P (1986) Characteristic-based schemes for the Euler equations. Annu Rev Fluid Mech 18:337–365. doi:10.1146/annurev.fl.18.010186.002005

Rubin H, Atkinson J (2001) Environmental fluid mechanics, 10th edn, Eastern hemisphere distribution. Marcel Dekker AG, New York

Vestas Wind Systems VG (2013) Vestas product brochures. In: www.vestas.com. http://www.vestas.com/en/media/brochures.aspx. Accessed 9 Dec 2013

Wells N (2012) Atmosphere–ocean interaction. In: The atmosphere and ocean: a physical introduction. Wiley, Chichester, pp 315–335

Chapter 6
Analysis 03: Future Scenario—OWF Development Within German EEZ

Political commitment regarding offshore wind energy supply asks for offshore demands in the German North Sea, precisely within Germany's exclusive economic zone (EEZ). There exist different scenarios for the EEZ utilization in the future, summarized in LOICZ report 2010 (Lange et al. 2010). At this juncture, the so-called scenario *B1* is the most interesting one for this study here. Scenario B1 defines the EEZ as an energy park separated into different stages of expansion of offshore wind turbines. An energy supply of 30 GW is politically planned till 2030, but that amount is not limited, and it cannot be excluded that further OWF expansion will be commissioned. Regarding the available areas within the EEZ, it is concentrated on one of the strongest possible realizable expansion called *B1-2030much*. Such an expansion shall supply around 90 GW of energy. The North Sea areas covered by wind turbines based on the scenario B1 and the expansion '2030much' is illustrated in Fig. 6.1. The specified area equates to 8,590 10-MW wind turbines in the atmosphere model METRAS, being set in a horizontal distance to each other by 1,990 m. Hence, the wind turbines are evenly spread over the OWF district.

This chapter shows the effect of such extreme amount of wind turbines in the North Sea and its dynamical and hydrographical conditions. Simulations are based on the setup North Sea simulations (TOS-02 in Sect. 3.3.2) and include two different ways of case studies. Case study I focuses on the effect of different wind directions; case study II considers the effect of OWFs under real meteorological conditions of June 2010. Anticipatorily, the model results based on expansion scenario *B1-2030much* does not simulate changes of the general North Sea circulation; that is why result presentation and analysis are focused on the area close to the OWFs where the OWF effects are identified (Fig. 6.1).

© Springer International Publishing Switzerland 2015
E. Ludewig, *On the Effect of Offshore Wind Farms on the Atmosphere and Ocean Dynamics*, Hamburg Studies on Maritime Affairs 31,
DOI 10.1007/978-3-319-08641-5_6

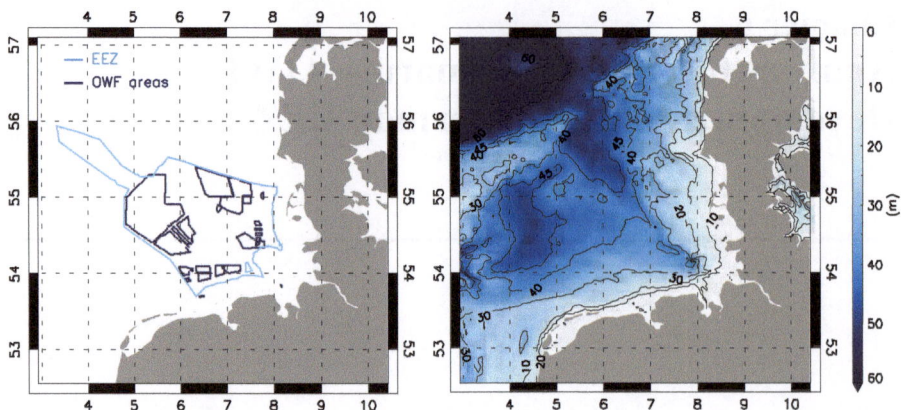

Fig. 6.1 *Left*: German EEZ (surrounded by *light blue lines*) and areas covered by wind turbines (surrounded by *dark blue lines*) based on scenario *B1-2030much*. Such expansion will supply 90 GW of energy, which equates to 8,590 wind turbines in the atmospheric model METRAS. *Right*: Bathymetry of German Bight in meters. Maximal depth in EEZ counts 60 m

6.1 Case Study I: Estimation of OWF's Impact by different Wind Directions

Case study I concentrates on the effect of OWF under scenario *B1-2030much* in matters of different wind directions. Under realistic meteorological conditions, wind speed and direction can vary with time, which again results in variations of ocean dynamics and can constrain constant formation of ocean dynamics. An approach of constant wind speed and direction is applied for better analyzing the possible effect the scenario *B1-2030much* can have on the German Bight. As mentioned, common wind direction within the German EEZ is southwest to west, but conditioned by the particular weather situation, other wind directions are possible. Thus, eight wind directions around wind rose are implemented, which are distinguished as N, SE, E, NE, S, SW, W, and NW. Simulation time for each wind direction counts 1 day. Again, runs without wind turbines and with operating wind turbines were necessary, and final results after 1 day of simulation are of interest.

6.1.1 Effect on the Atmosphere Over the German EEZ Based on Case Study I

This section provides atmospheric changes of the variables wind, temperature, and humidity over the German Bight simulated with METRAS for case study I. Figure 6.2 shows a representative collection of 10-m temperature, 10-m humidity

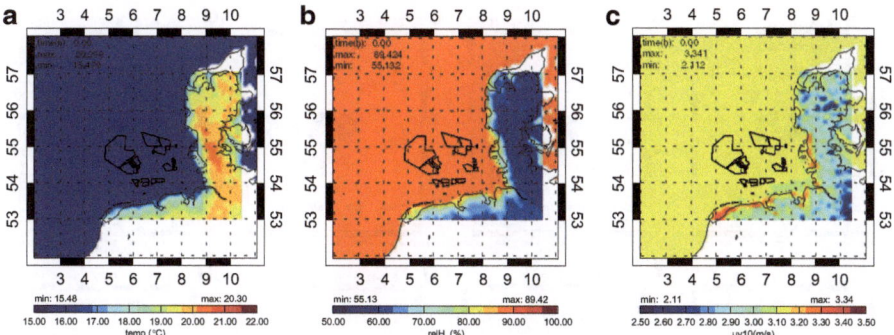

Fig. 6.2 Fields of 10-m (**a**) temperature, (**b**) humidity, and (**c**) wind speed after 1 day of METRAS simulation. Results belong to run with constant wind direction N as a representative example for all wind directions. Due to wind direction, the extrema can vary slightly; the important point is that in 10-m heights, the conditions are nearly homogenous over ocean. Outside of METRAS model area (Fig. 3.6), the METRAS data at boundaries were expanded over ocean

fields, and 10-m horizontal wind field of reference run without wind turbines. The important point of that representation is that over ocean the 10-m fields of temperature, humidity, and wind are homogenous. Hence, changes in those fields only occur due to operating OWFs.

After 24 h of simulation, the reference runs have a 10-m temperature of 14–15 °C over ocean, humidity is of 90 %, and wind speeds are of 3 m/s. Cut-in of wind turbines is set to 2.5 m/s, cut-off is set to 17.0 m/s in hub height. The wind field at hub height does not reach velocities greater than 17.0 m/s. Thus, it is assumed that the wind turbine parameterization is never avoided. In the following, differences between OWFr and REFr are presented for each defined wind direction after 1 day of simulation.

The changes of the horizontal wind velocity are shown in Fig. 6.3. Most of the EEZ area is influenced by a reduction of wind speed between 10 and 60 %. An intensified wake is formed especially within OWF areas. An increased wind speed of around 17 % up to 26 % occurs at the constraints of the wind farms depending on the wind direction. So also in the more realistic case of scenario *B1-2030much*, the structure of wake and wake's flanks can be identified. While the wind wake is strict locally limited over the OWF district, the wind increase also influences coasts and land depending on wind direction. In case of a wind coming from the coasts, that is, the south, southeast, and east directions, the wake length in wind direction is longer and broader over the ocean.

As mentioned in Chap. 4, here the use of the METRAS approach for wind wake simulation is necessary because, obviously, the Broström approach cannot cover such special formation for OWFs.

The changes of temperatures in 10 m are depicted in Fig. 6.4. Here, the temperature increases by about 3–5 % and partly also decreases by about 1 %–4 %. The rise in the temperature is especially located in lees of wind farm areas and also within OWFs and comprises a bigger zone than cooled areas.

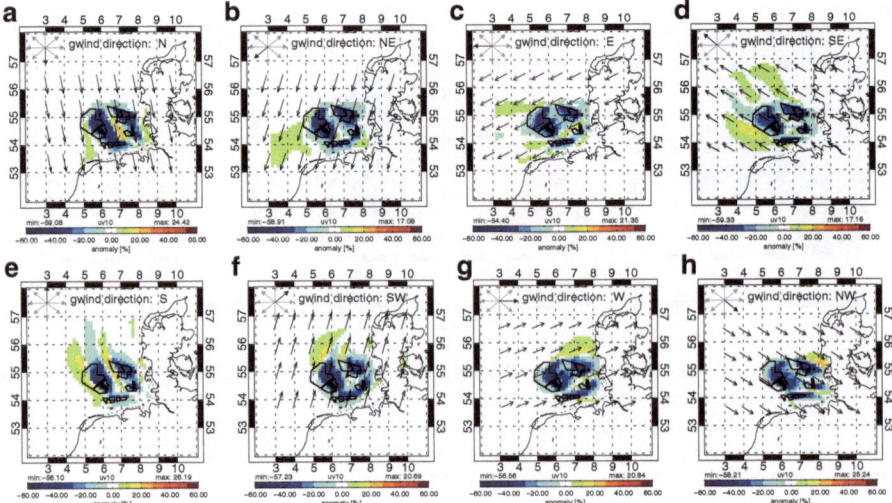

Fig. 6.3 Changes of 10-m horizontal wind field due to operating OWFs after 1 day of METRAS simulations for different wind direction cases. The change (OWFr–REFr) is given in percent. The prescribed constant wind directions at height of geostrophic wind are N (**a**), NE (**b**), E (**c**), SE (**d**), S (**e**), SW (**f**), W (**g**), and NW (**h**). Areas surrounded by *black solid lines* within the German Bight are OWF areas comprising 8,590 wind turbines. *Arrows* define real wind direction (OWFr) in 10-m heights. Maximal changes of 60 % are located at OWF districts. In sum, changes are regionally located within Germany's EEZ

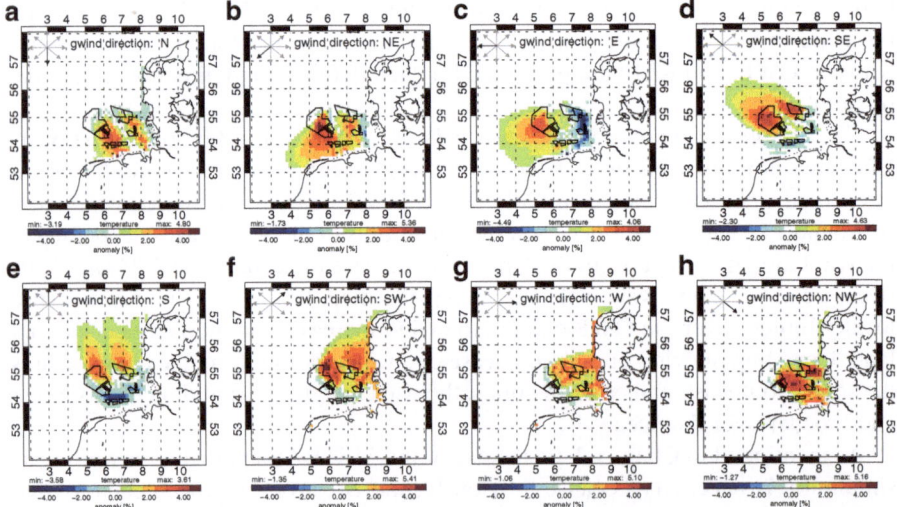

Fig. 6.4 Changes of 10-m temperature field due to operating OWFs after 1 day of METRAS simulations for different wind direction cases. The change (OWFr–REFr) is given in percent. The prescribed constant wind directions at height of geostrophic wind are N (**a**), NE (**b**), E (**c**), SE (**d**), S (**e**), SW (**f**), W (**g**), and NW (**h**). Areas surrounded by *black solid lines* within the German Bight are OWF areas comprising 8,590 wind turbines

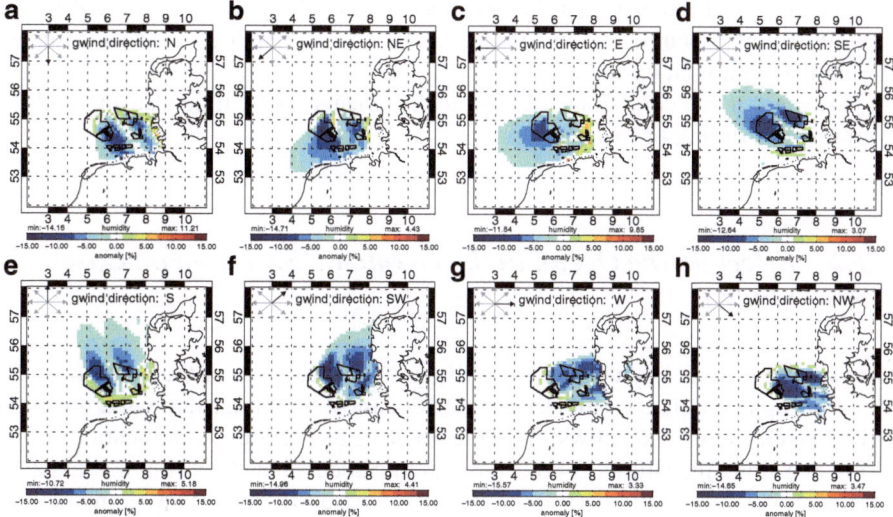

Fig. 6.5 Changes of 10-m relative humidity field due to operating OWFs after 1 day of METRAS simulations for different wind direction cases. The change (OWFr–REFr) is given in percent. The prescribed constant wind directions at height of geostrophic wind are N (**a**), NE (**b**), E (**c**), SE (**d**), S (**e**), SW (**f**), W (**g**), and NW (**h**). Areas surrounded by *black solid lines* within the German Bight are OWF areas comprising 8,590 wind turbines

A cooling is located windward of OWFs, that is, in wind direction in front of the OWFs and mostly connected with land. With the exception of case East wind, the temperature increase is slightly more dominant than the reduction.

The changes of the humidity in 10 m are shown in Fig. 6.5. The formation of the changes is similar to the changes of temperature. In case of warming, humidity is reduced by 11–15 %. In the case of cooling, humidity increases by 3–4 %, 11 % in the case of north wind and around 9 % in the case of east wind.

Precipitation and cloudiness were not formed within these simulation runs for all wind cases.

Over ocean, the OWFs normally lead to a cooling by around one degree (Linde et al. n.d.), while onshore farms lead to opposite effect (Baidya Roy 2004; Baidya Roy and Traiteur 2010; Zhou et al. 2012). Here, listed changes become quite constant after 8 h of simulation. Therefore, the warming cannot be caused by the diurnal cycle. The cooling and the increase of humidity are connected with warmer dryer air coming from land, which flows over ocean and advects moisture. The SST in METRAS is constant during the whole simulation time and is set to 15 °C. The 10-m temperature fields do not reach temperatures below that value. The warming in the area of wind reduction downstream of the OWFs is connected with vertical mixing and thus changes in the surface fluxes, as explained in Chap. 5 (Sect. 5.2.1). In reality, a more unstable stratification is expected during the night over water, supporting vertical mixing and a cooling in 10 m. During the day, the more stable stratification over water keeps the OWF induced cooling (Linde et al. n.d.). The overall cooling, documented in Linde et al. (n.d.), is a result of a stronger impact during the night than during the day.

However, here, conditions do not really change with time causing only a warming and drying of lower layers.

6.1.2   Effect on the Ocean in the German EEZ Based on Case Study I

This section provides dynamical and hydrographical changes of the ocean over the German Bight simulated with HAMSOM for case study I (constant wind direction over 1 day). Figure 6.6 shows the representative temperature stratification of the North Sea along the latitude 54.62° through the model area after 24 h of simulation of REFr based on conditions of June 2011. Also, an important point of that representation is that stratification over simulation time of 1 day does not vary in REFr. Hence, changes only occur due to the operating OWFs in OWFr. The maximal SSTs are around 13.0 °C in coastal areas, and at bottom temperature reaches 7.63 °C.

In the following, differences in fields between OWFr and REFr are presented for each defined wind direction after 1 day of simulation for surface elevation, vertical velocity component w, horizontal velocity components u and v, temperature, and salinity. As shown in the results of the surface elevation, the velocity components are based on runs with full meteorological forcing. The presentation of temperature and salinity is distinguished into wind and pressure forcing only and full meteorological forcing. That is necessary because here the OWF effect on the temperature indicates a warming, which is in opposition to the general OWF cooling over ocean that is found in literature. As analyzed in Sect. 5.3.5, the meteorological forcing including temperature and humidity mainly affects the hydrographic conditions of the ocean's upper layers, especially at the sea surface. The surface elevation and the

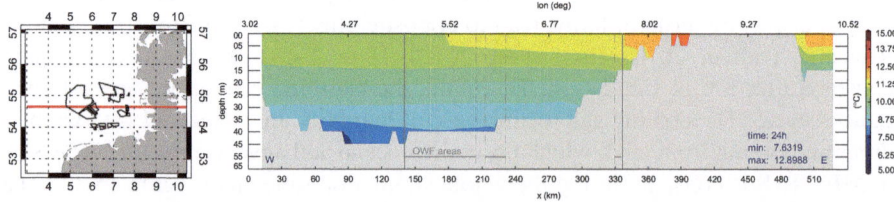

Fig. 6.6 *Left*: Latitude 54.62° (*red line*) through model area with scenario *B-2030much*. *Black* surrounded areas mark fields of offshore wind farms. *Right*: Reference temperature stratification along latitude of 54.62° based on June 2011 and SW wind direction as an example. The *gray shaded area* is land, respectively bathymetry. *Horizontal* and *vertical gray lines* mark the position of areas, including wind turbines. Values are given in degree Celsius. SST for other wind directions varies slightly in coastal regions

vertical velocity component are in average independent of the OWF effect on atmospheric temperature and humidity forcing fields in 10-m height. The horizontal velocity varies only in order of ± 0.001 m/s due to gradients in density fields triggered by determined temperature and salinity changes based on the forcing but do not affect vertical motion.

Changes in the surface elevation ζ in the case of scenario *B1-2030much* are illustrated in Fig. 6.8 for all eight wind direction cases. As the theoretical analysis implies, a dipole of ζ is even formed in the more realistic simulation of the German Bight. The magnitude of dipole's extrema is similar for each wind direction case within the range of -0.382–0.0308 m. Mostly, the minima results in a stronger effect than the increase, with the exception of wind direction cases SE, S, and SW. The positions of dipole's extrema depend on wind direction. Downstream behind and within the OWF (lee of OWF), often a minimal in ζ is identified and in front of the OWF, that is, in windward of the OWF, the maximum is detected. Changes in the surface elevation are connected with greatest wind stress and thus the wake in velocity components. Having u-component as the strongest content of wake, the wake areas lead to an increase of ζ; having v-component as the strongest wake, then the ζ-maximum occurs in the area of the v-component wake. That fits with the explanations of Ekman transport and divergence and convergence in Sect. 5.2.3.

Therefore, two positive extrema of the ζ-change exist in the case of wind direction S because over the OWF district, the v-component shows two areas of flow reduction; see Fig. 6.11. In the case of wind direction W, the u-component shows in Fig. 6.10 a wake over the OWF district with three local minima that end in three local minima at surface elevation between latitude $54°$ and $55°$. Although changes in surface elevation are in maximum order of 0.04 m, such a small change can have an economic relevance for the Elbe estuary concerning shipping and harbor industries. Here, changes of the surface elevation due to the OWF expansion should be considered besides tides in the case of bigger ships leaving and entering Elbe and Hamburg harbor because for these, a water-level change of a few centimeters plays an important role in order for them not to run aground. Such a change can also play a role in the case of storm surges.

The theoretical analysis in Chap. 5 underlines the formation of up- and downwelling cells due to the change in surface elevation. Such cells also occur in the German Bight (Figs. 6.7 and 6.9). Figure 6.9 illustrates the *change of the vertical velocity component w* at 12.5-m depth. In contrast to the two main vertical cells in theory, here, belts of up- and downwelling occur depending on the arrangement of wind farms within the EEZ.

The belts of maximal changes depend on surface elevation. In the case of a negative change in ζ, upwelling is simulated, and in the case of a positive change in ζ, downwelling is simulated, which agree with the theoretical results. The magnitude of vertical motion is independent of the wind direction due to a high agreement of extrema between wind direction cases. Maximal upwelling is 7.0×10^{-5} m/s (6.05 m/d), and the maximal downwelling has speeds of -1.0×10^{-4} m/s (8.64 m/d). In waters with depths of 30 m, such motion means an overturning within 3.5–5

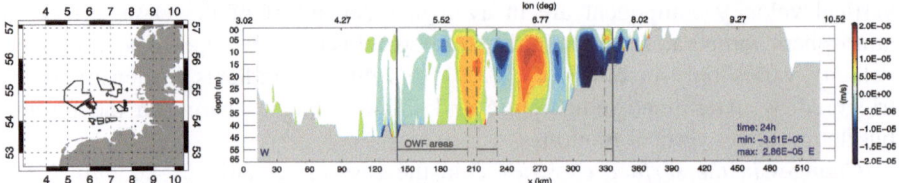

Fig. 6.7 *Left*: Latitude 54.62° (*red line*) through model area with scenario *B1_2030much*. *Black* surrounded areas mark fields of offshore wind farms. *Right*: Change of vertical velocity component *w* due to operating wind turbines along 54.62° for wind direction case SW as example to show that vertical cells affect whole ocean depth. *Gray shaded area* is land, respectively bathymetry. *Horizontal* and *vertical gray lines* mark the position of areas, including wind turbines. Values are given in m/s. Here, minimal/maximal vertical motion equates to −3.11/2.47 m/d

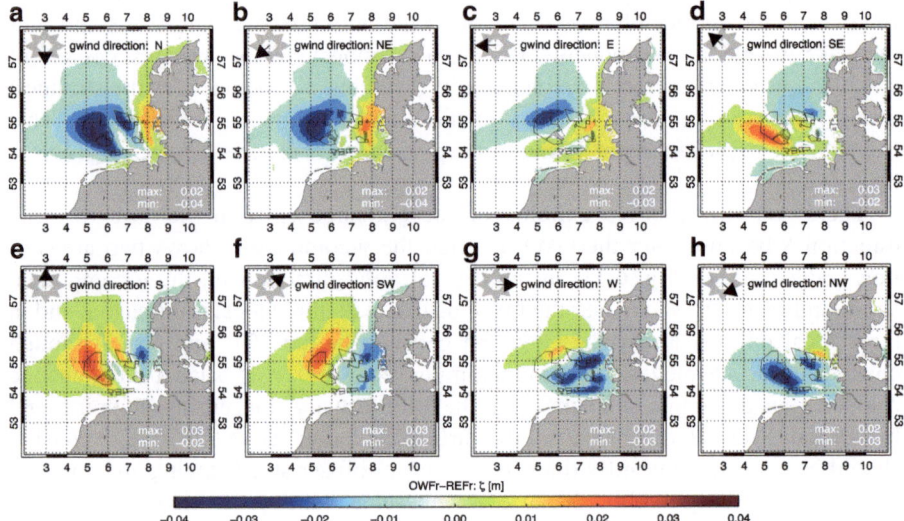

Fig. 6.8 Change in surface elevation due to OWFs in case of different wind direction cases (gwind direction (**a**)–(**h**) from N to NW) after 1 day of simulation. The wind direction is defined at height of geostrophic wind. Units given in m. *Dark gray shaded area* marks land; *black lines* illustrate OWF districts. Results are for full forcing

days. Figure 6.7 specifically exemplifies in the case of wind direction SW that the vertical motion affects the whole model depths with maximal changes between 10 m and 15 m. The intensity of the OWF induced vertical velocity *w* is stronger over the whole depth towards the coasts, especially for downwelling. The analysis of the OWF effect on the ocean system in Sect. 5.3.6 shows that intensification is supported by shallower water.

Changes of the velocity components u and v at surface due to OWFs are illustrated in Figs. 6.10 and 6.11. Depending on the wind direction the velocity

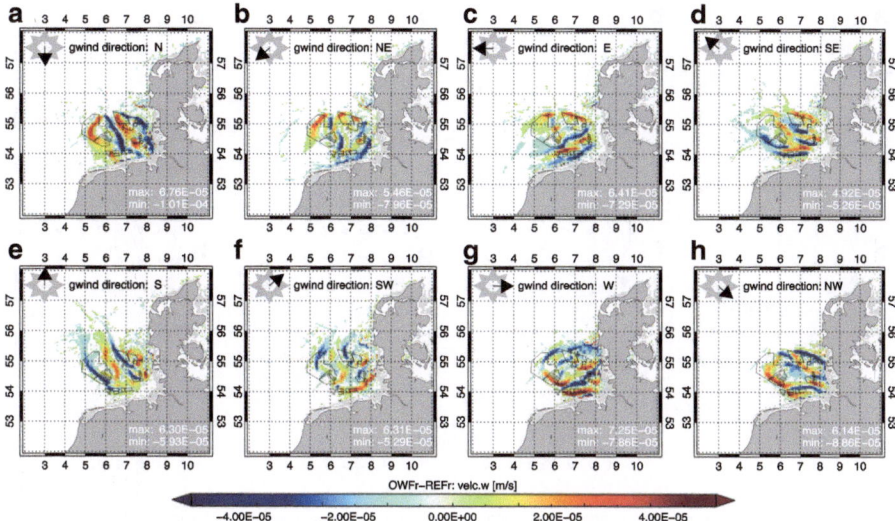

Fig. 6.9 Change in velocity component *w* at 12.5-m depth due to OWFs in case of different wind direction cases (gwind direction (**a**)–(**h**) from N to NW) after 1 day of simulation. The wind direction is defined at height of geostrophic wind. Units given in m/s. *Dark gray shaded area* marks land, *light gray shaded area* marks bathymetry, and *black solid lines* illustrate OWF districts. Minimum counts −6.05 m/d, maximum 8.64 m/d, which means an overturning after 3.5–4.9 days in areas of 30 m of water depth. Results are for full forcing

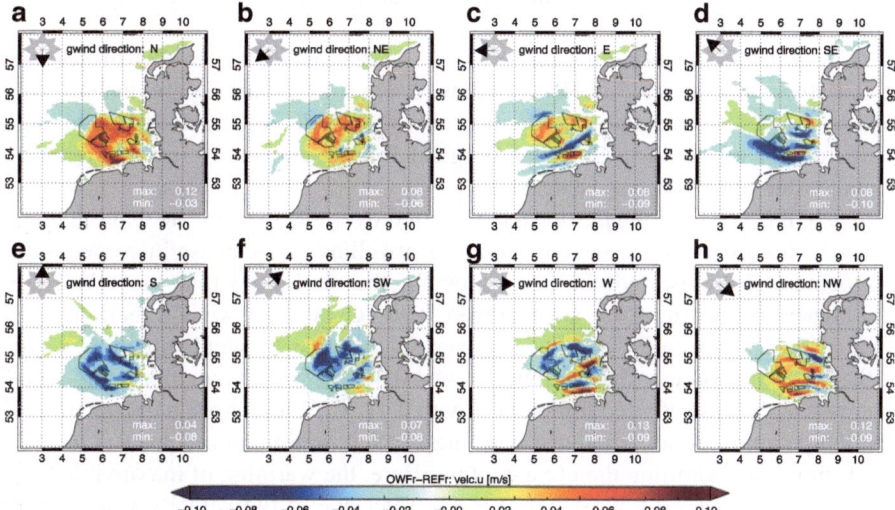

Fig. 6.10 Change in velocity component *u* at surface due to OWFs in case of different wind direction cases (gwind direction (**a**)–(**h**) from N to NW) after 1 day of simulation. The wind direction is defined at height of geostrophic wind. Units given in m/s. *Dark gray shaded area* marks land; *black lines* illustrate OWF districts. Results are for full forcing

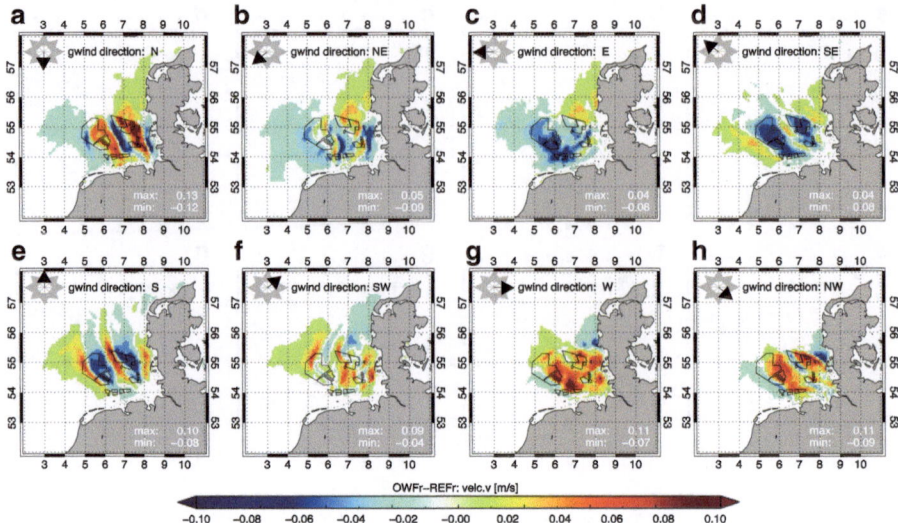

Fig. 6.11 Change in velocity component *v* at surface due to OWFs in case of different wind direction cases (gwind direction (**a**)–(**h**) from N to NW) after 1 day of simulation. The wind direction is defined at height of geostrophic wind. Units given in m/s. *Dark gray shaded area* marks land; *black lines* illustrate OWF districts. Results are for full forcing

wake is stronger dominated by the *u*-component respectively by the *v*-component. At surface the changes are in order of ± 0.10 m/s, which means an increase/ decrease of 20 % compared to the reference run with average horizontal velocities at surface of 0.5 m/s.

The effect in temperature and salinity fields due to the OWFs in scenario *B1-2030much* is depicted in Figs. 6.12, 6.13 and 6.14 at surface and in Figs. 6.13 and 6.15 at 12.5-m depths. The figures show the results for the ocean simulations with forcing neglecting full meteorological forcing and clarifying hydrographic changes due to dynamical changes.

The OWF effect on temperature is scattered over the areas of the OWF induced vertical motion. In case of upwelling/downwelling, a decrease/increase of the temperature is registered. In 12.5-m depth, the cooling within the OWF area dominates the warming by an averaged 0.36 °C over all wind direction cases, Fig. 6.12. The strongest cooling is given in the case of wind direction S with -0.42 °C. On average, the warming counts 0.30 °C, with a maximal temperature increase in the case of wind direction N with 0.41 °C. While in the depth changes in the temperature are in coherence with the vertical motion, at sea surface the SST leads more to a warming than to a cooling. Here, the warming of maximal 0.24 °C temperature (in case of West wind) is an effect due to the velocity wake, which shifts the temperature front. That effect is pointed out in Fig. 6.16, which exemplifies wind direction N. SSTs are shown for reference run REFr and run with operating wind turbines OWFr separated for simulation with wind and pressure

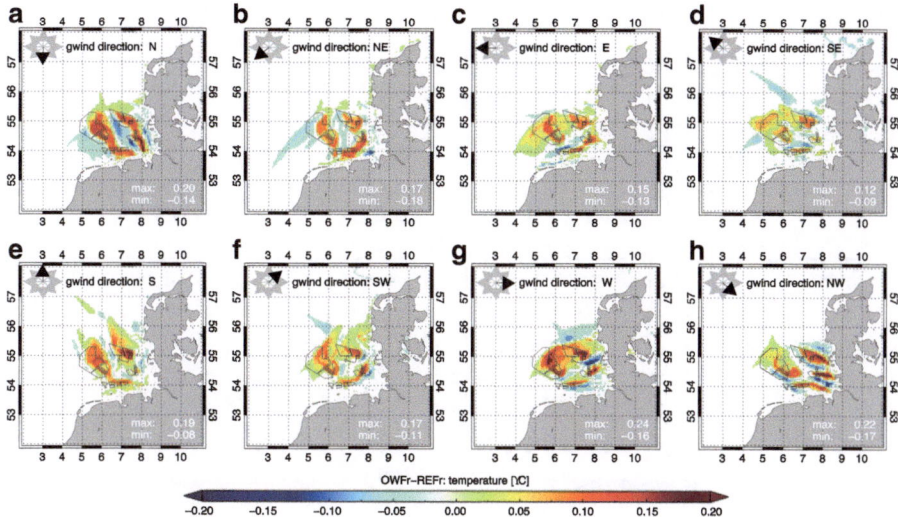

Fig. 6.12 Change in SST due to OWFs in case of different wind direction cases (gwind direction (**a**)–(**h**) from N to NW) after 1 day of simulation. The wind direction is defined at height of geostrophic wind. *Dark gray shaded area* marks land; *black lines* illustrate OWF districts. Results are only for wind and pressure forcing

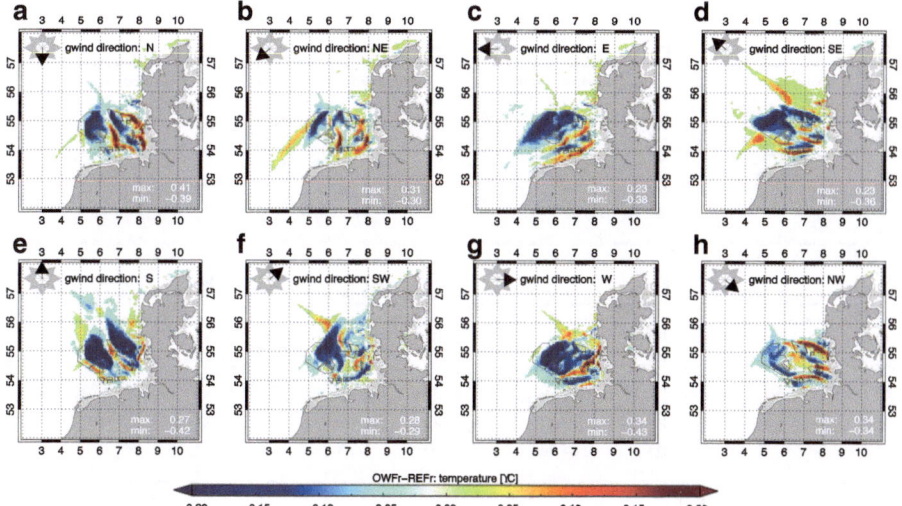

Fig. 6.13 Change in temperature at 12.5-m depth due to OWFs in case of different wind direction cases (gwind direction (**a**)–(**h**) from N to NW) after 1 day of simulation. The wind direction is defined at height of geostrophic wind. Units are given in degree Celsius. *Dark gray shaded area* marks land, *light gray shaded area* marks bathymetry, and *black solid lines* illustrate OWF districts. Results are only for wind and pressure forcing

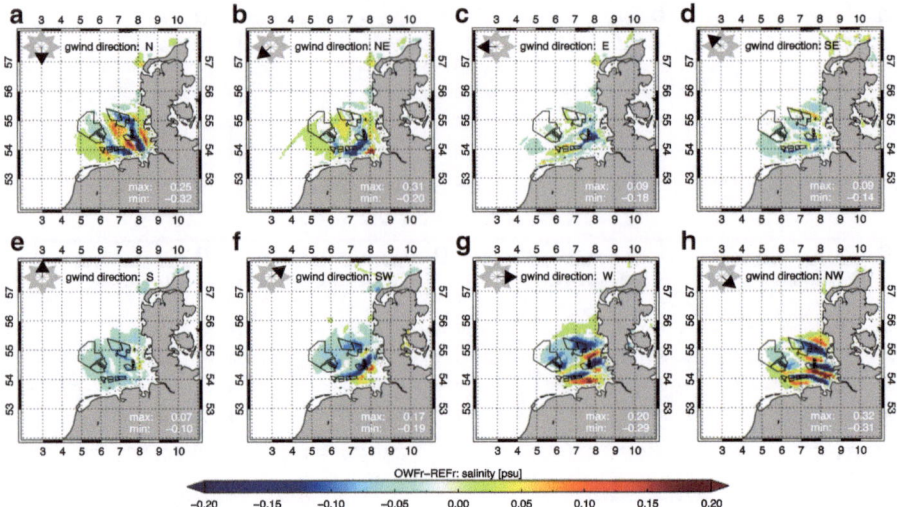

Fig. 6.14 Change in salinity concentration at surface due to OWFs in case of different wind direction cases (gwind direction (**a**)–(**h**) from N to NW) after 1 day of simulation. The wind direction is defined at height of geostrophic wind. Units are given in psu. *Dark gray shaded area* marks land; *light gray shaded area* marks bathymetry. Results are only for wind and pressure forcing

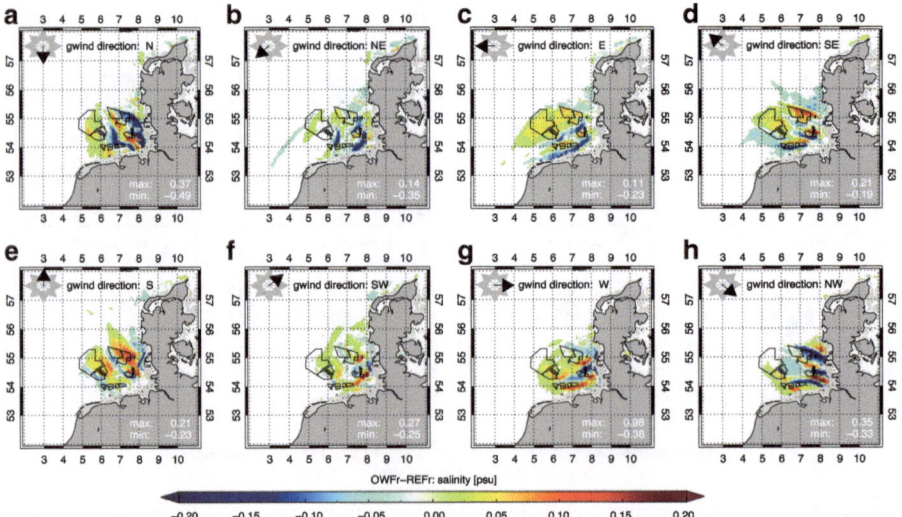

Fig. 6.15 Change in salinity concentration at 12-m depth due to OWFs in case of different wind direction cases (gwind direction (**a**)–(**h**) from N to NW) after 1 day of simulation. The wind direction is defined at height of geostrophic wind. Units are given in psu. *Dark gray shaded area* marks land, *light gray shaded area* marks bathymetry, and *black solid lines* illustrate OWF districts. Results are only for wind and pressure forcing

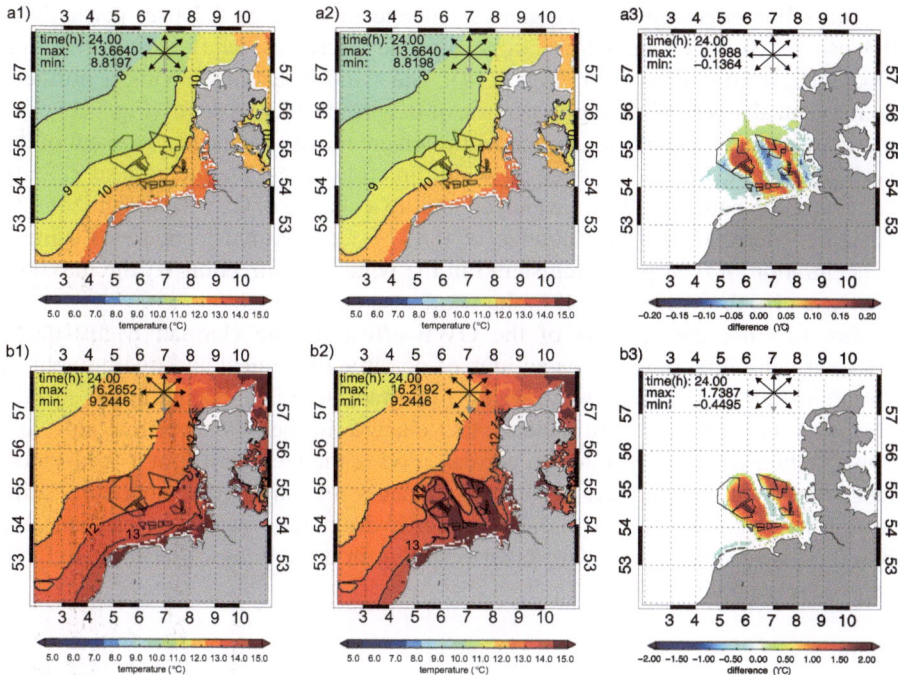

Fig. 6.16 SST within the German Bight for REFr (a1,b1) and OWFr (a2,b2) and differences between OWFr and REFr (a3,b3). *Upper row* presents results for SST in case of wind and pressure forcing; *second row* pictures SSTs in case of unrestrained forcing. Results show exemplary OWF effect on SST for wind direction case N wind. OWF affects ocean flow, which leads to shift in temperature front causing warming/cooling at surface compared to reference run

forcing only and full meteorological forcing after 1 day. The differences between runs OWFr and REFr in SST rely on a deformation of the temperature front, visible at the 9 and 10 °C contour line, which is deformed by the velocity wake. The shift in the case of full forcing is superposed by the impact on the temperature forcing on SST.

While dynamics only end in a temperature effect, due to the OWF by ±0.20 °C, the full meteorological forcing leads to an average SST warming of 1.68 °C and a cooling of 0.72 °C. Considering the full forcing and the OWF induced warming at 10-m heights, the SST and the upper ocean are significantly warmer than in the case of wind and pressure forcing only. A wind and pressure forcing only represents only 12–29 % of the OWF effect on ocean temperature related to simulation results under full forcing. But considering now that under more realistic meteorological conditions the atmosphere will cool, as mentioned before, and bearing in mind that simulations show dominantly an upwelling in the OWF district, it must be expected that the German Bight would suffer an overall cooling.

The OWF effect on salinity gives a change in order of 0.2 psu, particularly 0.3 psu. At surface, the salinity concentration decreases for the wind direction clockwise from east to west (Fig. 6.14). The wind directions having a northern component also result in a slight increase of 0.3 psu. In 12.5-m depth (Fig. 6.15), the increase and decrease of salinity concentration are mostly balanced as a result of the vertical motion. In average over all wind direction cases, the increase of salinity concentration counts 0.33 psu, which equates to a change of 0.95 %, and the salinity decrease counts 0.31 psu, a change of 0.88 %. A maximal increase of salinity concentration of 0.98 psu is registered in the wind case direction W, the maximal decrease of −0.49 psu is found in the wind case direction N.

Summarizing the analysis of the OWF effect on the German Bight, under constant wind directions, gives an idea of possible dynamical and hydrographical variations due to produced wind wake. Important here is a change in the surface elevation in order of centimeters; the change in belts of vertical up- and downwelling is in order of 3–5 m per day, giving the German Bight a 'whirlpool' character. Hydrographic changes are significantly related to the temperature/salinity in order of ± 0.2 °C and ± 0.3 psu. But hydrographic variations can increase based on atmospheric boundary layer resulting in a warming of more than 1 °C, respectively cooling.

6.2 Case Study II: OWF's Impact Based on a Real Meteorological Situation

Various theoretical assumptions were analyzed to estimate OWFs' effect on the atmosphere and, especially, on the ocean. For a better estimation of the OWF effect under daily wind variability, the simulations of scenario *B1-2030much* consider realistic meteorological conditions. Therefore, a meteorological situation in June 2010 is chosen as a realistic example.

Simulations of June 2010 were done over the days 16–19 again for operating wind turbines (OWFr) and for no wind turbines (REFr). The wind turbines only operate at wind speeds between 2.5 and 17 m/s at hub height. The ocean runs are based on North Sea simulations (TOS-02), described in Sect. 3.3.2. The result presentation focuses on daily means calculated by a 10-min mean model output.

The mid of June 2010 denotes an interesting weather situation with a strong cooling effect over Europe, with strong precipitation events, including snow in the Alps. Synoptic inspection shows that Germany was in sphere of a long-wave trough with an expansion from Scandinavia up to the Mediterranean Sea. The focus of the activity was placed on Scandinavia, which was connected with a ground low pressure over mid-Scandinavia. During the days from 16 to 19 June 2010, the frontal system crossed Germany from NW to SE, which results, especially over the Alps, in extensive rainfall. To easily integrate that meteorological situation over the 4 days, Fig. 6.17 shows 500 hPa geopotential for each day.

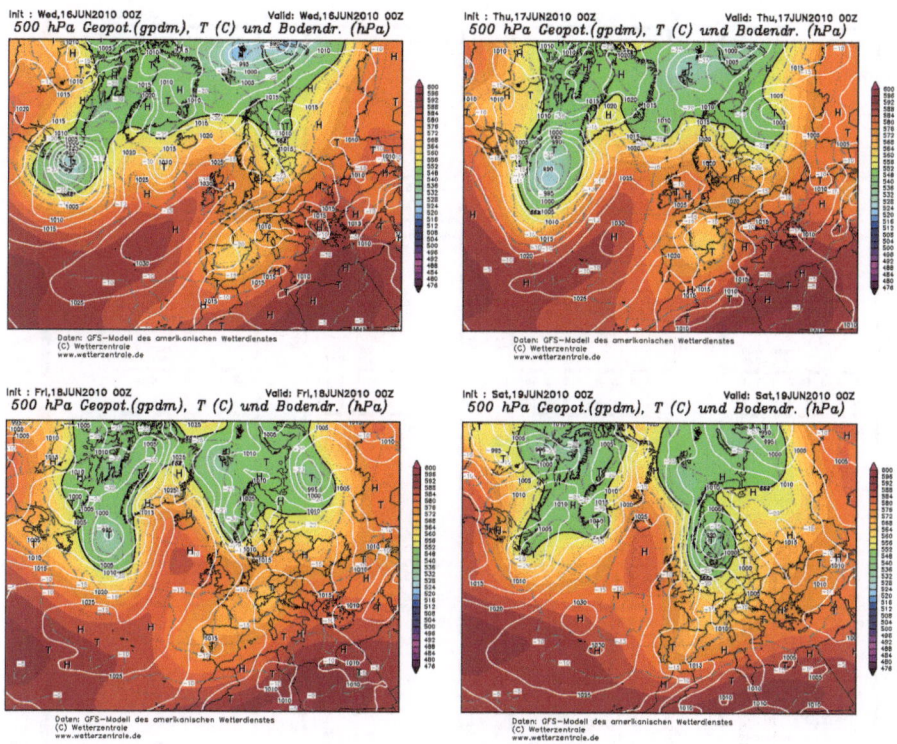

Fig. 6.17 500 hPa geopotential from 16 to 19 June 2010. Over the 4 days, a long-wave trough affects the meteorological situation of northern Europe and Germany. Maps are taken from www. wetterzentrale.de

6.2.1   Effect on the Atmosphere Based on Case Study II

The mean *wind direction* over the German Bight in 10-m height was NE to N. Figure 6.18 summarizes the effect of OWFs based on scenario *B1-2030much*:

The *formation of the wind wake* is similar to the wind direction case N and NE described in previous section. The daily means of 10-m wind fields are more or less equal for day 16th and 17th with wind speeds of 10 m/s. Here, the maximal wind reduction of 5.35 m/s and an increase of 1.40 m/s occur especially towards coasts. Only the daily mean of the wind direction slightly varies between those 2 days. On 18th of June 2010, the OWF effect is weaker due to wind speeds of less than 10 m/s over the German Bight. But on 19 June, the wind reduction counts up to 9.03 m/s. The 10-m wind speeds on that day simulated in run OWFr are in average 11 m/s. The wind turbines operate over the whole time of simulation.

The effect of the OWFs on the *10-m temperature field* obviously shows the previously mentioned generally detected cooling of around 0.5 °C over ocean.

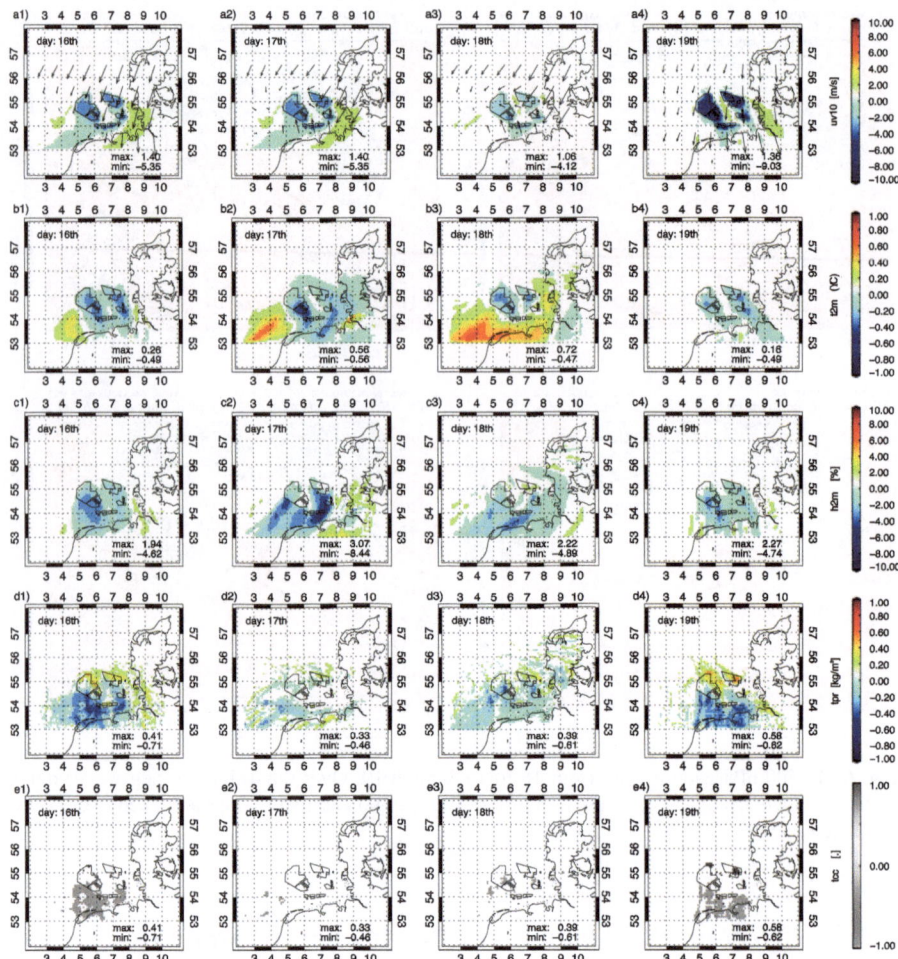

Fig. 6.18 Daily mean change (OWFr–REFr) of meteorological forcing due to OWF for 16th (**a1–e1**), 17th (**a2–e2**), 18th (**a3–e3**) and 19th (**a4–e4**) of June 2010. From top to bottom: 10-m horizontal wind field (**a1–a4**), 10-m temperature (**b1–b4**), 10-m humidity (**c1–c4**), total precipitation (**d1–d4**), and presence of clouds (**e1–e4**). Cloudiness means *dark gray*—(1) OWFr has clouds, and REFr has no clouds; *light gray* (−1) means OWFr has no cloud but REFr has clouds; *white* (0) means the presence of clouds is similar for OWFr and REFr. *Arrows* in the figures of the horizontal wind field show wind direction of OWFr. *Solid black lines* illustrate OWF districts of expansion scenario B1-2030much

Particularly, a warming is identified with maximal temperature of 0.72 °C at day 18 connected with offshore wind. On 19th of June, the cooling dominates due to a strong wind wake. The effect of the OWFs on the *10-m humidity field* consists of a drying of around 5 % and a reduction of total *precipitation* by an average of 0.005 kg/kg. Due to such effect, in case of run OWFr, the *presence of clouds*

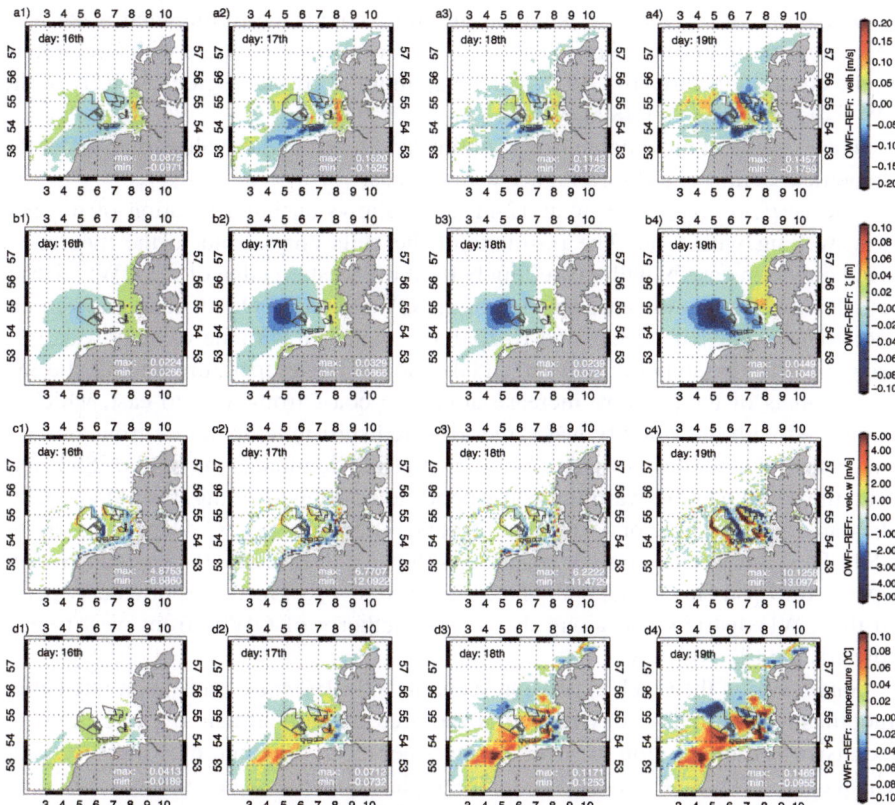

Fig. 6.19 Daily mean change (OWFr–REFr) of ocean conditions due to OWF for 16th (**a1–e1**), 17th (**a2–e2**), 18th (**a3–e3**), and 19th (**a4–e4**) of June 2010. From top to bottom: Horizontal velocity field at surface (**a1–a4**), surface elevation (**b1–b4**), vertical velocity component w at 12.5-m depth (**c1–c4**), SST (**d1–d4**), and salinity concentration at surface (**e1–e4**). *Dark gray shaded areas* are land, *light gray* (at velocity w) marks bathymetry, and *solid black lines* illustrate OWF districts of expansion scenario *B1-2030much*

southerly of the wind farm areas and northerly of the German coast on 16 and 19 June 2010 was not detected, like in REFr. On the other hand, OWFr leads to some cloudy spaces within the OWF district on 19 June northerly of latitude 54.00°, which are not given in REFr.

6.2.2   Effect on the German Bight Based on Case Study II

Even for the ocean simulations, results for days in June 2010 are comparable with the theoretical approach of wind direction N/NE in previous section. The daily means of the OWF effect on the ocean on June 2010 are documented in Fig. 6.19:

The *horizontal velocity* wake areas are more connected to the wind wake with reductions in order of 0.15 m/s. The *surface elevation* has a dipole with a minimum in the ocean in the area of OWFs and a maximum in coastal regions easterly of the OWF districts of −0.1048 and 0.0449 m.

The extrema of the dipole grow with time and reach higher values than under the theoretical conditions due to higher wind speeds.

The *vertical component w* at 12-m depth shows an intensification of up- and downwelling with time. The positions of the belts correspond to the theoretical wind direction case N with upwelling belts at the western edge of the OWF district and downwelling at the eastern edge of the OWF district. On 19 June, the vertical motion has values of −13.10 and 10.13 m/d. Connected with the velocity wake, the triggered shift in temperature front, and the increase in temperature forcing, the *SST* shows areas of temperature increase along the coast from west to east, which is connected with the velocity wake, the triggered shift in temperature front, and the increase in temperature forcing. Again, that effect dominates the field of SST, and with the depth cooling/warming corresponds to up-/downwelling. On 19th of June, the daily mean change of SST counts around +1.5 and −0.1 °C. Depending on the SST, the *salinity* concentration at surface shows a decrease of −0.18 psu on 16 June up to −0.79 psu on 19 June and an increase of 0.77 psu. Decreases obviously occur in the OWF districts and along the coasts southerly of OWFs from west to east towards the Elbe estuary.

Summarizing the OWF expansion scenario *B1-2030much* leads to an intensified modification of the North Sea within the area of the German Bight. Due to an extensive vertical motion of several meters per days, triggered by wind speed and change in surface elevation, the hydrographic conditions are strongly affected. Special consideration to the development of the OWF effect on the ocean must be borne in mind. Depending on the wind direction and the wind speed, ocean conditions will easily vary within the German Bight due to OWFs with currently unknown and unassessable effect on the ecosystem.

References

Baidya Roy S (2004) Can large wind farms affect local meteorology? J Geophys Res 109:D19101. doi:10.1029/2004JD004763

Baidya Roy S, Traiteur JJ (2010) Impacts of wind farms on surface air temperatures. Proc Natl Acad Sci U S A 107:17899–17904. doi:10.1073/pnas.1000493107

Lange M, Burkhard B, Garthe S, Gee K (2010) Analyzing coastal and marine changes: offshore wind farming as a case study. LOICZ Res Stud 36:212

Linde M, Hoffmann P, Lenhart HJ, Schlünzen KH. Influence of large offshore wind farms on urban climate 1–1

Zhou L, Tian Y, Roy SB et al (2012) Impacts of wind farms on land surface temperature. Nat Clim Chang 2:1–5. doi:10.1038/nclimate1505

Chapter 7
Summary, Conclusion, and Outlook

With the objective of the evaluation of the influences that offshore wind farms can possibly have on the atmosphere and the ocean, this study deals with the analysis of the physical OWF effect. At this juncture, model simulations were consulted and set in relation to measurements. Thereby, one focus is the physical, theoretical coverage of the OWF effect, and the other aim is the determination of, especially, oceanic changes in the future of the North Sea within the German Bight due to politically planned OWF expansion.

The theoretical analysis of the OWF effect on the atmosphere shows that wind turbines produce a wind wake downstream of wind farms. Depending on wind speed, the amount of turbines, their arrangement, and spanned area, the wind wake consists of a maximal simulated wind speed reduction of 70 % as stronger wind is as intensive as the wake. The wake's maximum is located within the wind farms, and behind the wind farms wind again increases with distance to OWF center. A change from the operating mode to nonoperating mode of the wind turbines result in advection of the provoked wind wake by the mean wind. Depending on atmospheric conditions and surface fluxes, an OWF can cool or warm an atmospheric boundary by ±1 K, which also affects humidity. The dimension of the wake tail counts more than 120 km downstream of the OWF, and the wake dimension orthogonal to wind direction is determined by OWF configurations.

The description used here of the wake is based on two approaches: the approach by Broström and the wind turbine parameterization of atmospheric model METRAS. METRAS leads to more realistic results when it comes to satellite data compared to the Broström approach and has more advantages in matters of turbine types and atmospheric simulations. Hence, results of the OWF effect on the atmosphere are based on METRAS simulations. A technical simplification of the wake description can be the use of statistical description and analysis of the OWF induced variations of the atmosphere, like the example analysis of the impact of wind direction on the wake over the German Bight, placed in Chap. 6. Only model

© Springer International Publishing Switzerland 2015
E. Ludewig, *On the Effect of Offshore Wind Farms on the Atmosphere and Ocean Dynamics*, Hamburg Studies on Maritime Affairs 31,
DOI 10.1007/978-3-319-08641-5_7

simulations can develop such statistics, but they can be applied to ocean simulation by manipulating forcing fields due to OWF effect statistics.

The consequences in atmosphere due to operating OWFs significantly affect the ocean system. The wind wake causes a wake in the ocean velocity field within and behind the wind farm, which is connected with a reduced Ekman transport causing divergence and convergence of the water masses. That ends in a change of the surface elevation and the barotropic pressure field in the form of a dipole structure having an increase of surface elevation north to the wind wake and a decrease south to the wind wake. These effects on the ocean surface again cause vertical motion in order of several meters per day. Vertical motions mean cells around the OWF of upwelling and downwelling advecting the temperature field, which results in an excursion of the thermocline around the OWF in the vertical of possible 10 m, depending on the ocean's stratification. Respectively, the salinity and density fields are affected.

Analyses of external impacts triggering the OWF effect on the ocean systems lead to the result that besides the ocean depth, primarily, the wind wake defines the intensity and dimension of the OWF induced effect on the ocean system:

- Shallower waters intensify the up- and downwelling cells in the vertical, and hence stronger hydrographic changes are detected at the depth of the thermocline.
- A more intense wind wake leads to greater magnitudes of the up- and downwelling cells and stronger changes in the hydrography. Additionally, the vertical exclusion of the thermocline increases with the intensity of the wind wake, as well as the depth of the thermocline towards lower ocean layers.
- A wider wind wake (orthogonal to the wind direction) leads to a greater horizontal dimension of the velocity wake, which triggers the horizontal dimension of vertical cells and of the thermocline exclusion. The effect is positively linked, so a greater wake results in a greater horizontal OWF effect in the ocean.

The wind wake itself, and so the wake in the ocean velocity field, is defined dominantly by wind speed; by the number of wind turbines of an OWF, respectively the number of wind turbines within one model grid cell; and by the OWF's size, respectively the number of grid cells comprising the OWF district. Here, the wind wake follows the relation that the stronger the wind speed within the range of OWF's operation mode, the stronger the wind wake; the higher is the amount of wind turbines within one grid cell, the more intense is the wind wake; and the larger the OWF district with operating wind turbines is, the lager is the affected area of extreme wind reduction.

Additionally, the choice of the ocean forcing influences the OWF effect. The use of a full meteorological forcing for ocean simulations shows that the influence of temperature and humidity forcing is the OWF effect on the SST field positively overlaid. But the influence of offshore wind farms is dominantly driven by the 10-m wind forcing field.

The OWF impact on the ocean system occurs within minutes after switching to the operational OWF mode. Simulations show that the duration of the OWF effect lasted over 2 days after turning off the wind turbines, but due to the appearance of inertial oscillation in model runs, the OWF effect on the ocean system can vanish faster.

Evaluation with model results, taken around the test wind farm alpha ventus, shows a good agreement of the OWF effect on the ocean's temperature stratification. The model restrictions only lead to expected discrepancies in the horizontal, but the main simulated phenomena of up- and downwelling cells causing changes in the temperature stratification are also established in nature.

Simulations over the German Bight underline that detected influences of OWFs, presented under theoretical assumptions, show that the impact can be intensified in reality. Addicted to meteorological situation and conditions of boundary layer, OWFs mostly have a cooling effect under realistic conditions. Instead of the vertical cells, belts of vertical motion, with maxima of ± 10 m/d, occur depending on wind direction, which controls the establishment of the surface elevation dipole structure. The OWF induced atmospheric cooling, respectively warming, is stamped on SST, which is also changed by an identified switch of the SST front due to the wake in the ocean's horizontal velocity field. Theoretically, SST can vary due to the operating wind turbines by ± 1 °C.

Finally, such imposed dynamical and hydrographic modifications on the ocean system by OWFs yield to questions about possible variations on the ocean ecosystem and especially about biological consequences.

The theoretical approach of the dynamical analysis, influenced by OWF results, can be transferred to each offshore wind farm location in the world. Technically, it is possible to build OWFs along all coasts around the world. Independent of the habitat, physical up- and downwelling will occur. And there are plans all over the world to use offshore wind energy. So assuming that installed offshore wind farms will consistently operate, the described influence of OWF on the atmosphere and, especially, on the ocean will be of permanent duration in the future. The meteorological warming and cooling, as well as cloud dissipation and fog production, can affect local climates over time.

Regarding the ocean's effect, the surface elevation can impact shipping and storm surges, but the important OWF effect on the ocean is the occurrence of vertical motion, and so the influence on the ocean's stratification will be dominant. A warming or cooling of the upper layers and mixing over the whole water depth will have consequences on the ecosystem. The response of the ecosystem to the vertical mixing regarding nutrients, plankton, and other microorganism depends on the habitat. But the OWF induced upwelling, in comparison with the coastal up- and downwelling, is special because water that carries the properties of the ecosystem will have relevant consequences on the mixed layer, as mentioned already by Broström.

Additionally, an impact on mammals is possible. For example, the simulated up- and downwelling belts in the North Sea can trigger distribution of common porpoise (personal correspondence with Michael Dähne).

Hence, the OWF induced dynamical changes of the ocean on biology, as well as on chemistry, should be a next step to estimating the OWF impacts on the ecosystem.

But besides the question about consequences for the ecosystem, the back coupling of the ocean with the atmosphere should be considered in the future as well. A change in SST of $1/2°$ again affects the atmospheric boundary layer and the atmosphere–ocean interaction. Analysis of OWF impacts on local climates, as well as impacts on the ocean, will need fully coupled atmospheric–ocean simulation to judge the strength of possible changes due to offshore wind farming.

Chapter 8
Appendix I: List of Data and Personal Correspondences

8.1 Personal Correspondence

Marita Linde
Meteorologisches Institut der Universität Hamburg
Bundesstrasse 55
20146 Hamburg
Germany

Michael Dähne
Institute for Terrestrial and Aquatic Wildlife Research (ITAW)
University of Veterinary Medicine Hannover (Foundation)
Werftstraße 6
25761 Büsum
Germany

8.2 Data Overview

ECMWF ERA-Interim data used in this work have been obtained from the ECMWF data server http://www.ecmwf.int/products/data, and additional ECMWF forcing data are a courtesy of the Meteorological Institute of the University of Hamburg.

METRAS data used in this work were simulated by M. Linde and are a courtesy of the Meteorological Institute of the University of Hamburg.

HAMSOM data used in this work were simulated by myself.

© Springer International Publishing Switzerland 2015   149
E. Ludewig, *On the Effect of Offshore Wind Farms on the Atmosphere and Ocean Dynamics*, Hamburg Studies on Maritime Affairs 31,
DOI 10.1007/978-3-319-08641-5_8

WOA-01 data used in this work are a courtesy of the Institute of Oceanography of the University of Hamburg.

ADCP and CTD data used in this work were supported by the BSH and measured by myself.

Chapter 9
Appendix II: Addition to WEGA Cruise and Result Presentation

9.1 WEGA Cruise 141

9.1.1 Impressions of WEGA Cruise 141

Due to security instructions, offshore wind turbines cannot be easily observed from close range. With the exception of people involved in this wind business, nobody else has really the opportunity to visit such an offshore wind farm. Indeed, there exist some tours, for example, from Helgoland going close to the German's test wind park *alpha ventus*. To get a feeling of the dimension of wind turbines, here, some pictures photographed during the "WEGA Cruise in May 2013" helps to see what happened outside of the North Sea. It is impressive how big, even smaller, offshore wind turbines are, compared to onshore wind turbines, which have a rotor diameter of 200 m and may have more in the future. On the other hand, it shows a proud technique, designed for more than 15 years of challenges against salty water and air, gears, and high swell.

At this point, I would like to take again the opportunity to thank the BSH for its special support.

Figure 9.1 shows VWFS WEGA; Figs. 9.2, 9.3, 9.4, and 9.5 illustrate the *alpha ventus* with wind turbines, research platform, Fino1, and relay station.

9.1.2 CTD Probe

Figure 9.6 shows CTD (conductivity, temperature, depth) measuring station at research vessel WEGA with winch and CTD probe and sensing elements. A CTD probe determines conductivity, temperature, and pressure of water, and based on that, salinity and density can be calculated.

© Springer International Publishing Switzerland 2015
E. Ludewig, *On the Effect of Offshore Wind Farms on the Atmosphere and Ocean Dynamics*, Hamburg Studies on Maritime Affairs 31,
DOI 10.1007/978-3-319-08641-5_9

Fig. 9.1 VWFS WEGA, sounding, wreck searching, and research vessel of the BSH. Photo by BSH

Fig. 9.2 Impression of the German test wind park alpha ventus on 12 May 2013 in the morning (*left*) and in the evening (*right*)

Temperature Measurement

A platinum thermometer Pt100 is commonly used for temperature measurement via CTD. That is a resistance wire, which has a temperature of 0 °C and an electrical resistance of 100 Ω. Platinum is used due to its long-term stability and reproductibility of electrical conditions. The coefficient of specific resistance is positive, which means resistance grows with temperature. The relationship is linear, but the temperature coefficient of platinum is small, $3.85 \times 10^{-3}/°C$. The level of measured accuracy requires a value of 0.01 °C; the aimed resolution is 0.001 °C.

Fig. 9.3 Research platform Fino1 (*left* and *middle*) close to alpha ventus and relay/transformer station close to alpha ventus wind turbine AV12 (*right*), taken on 12/13 May 2013

Fig. 9.4 Impressions of dimension of offshore wind turbine and nacelle. Helicopters bring people to offshore wind farm and let them down on a rope for maintenance work at wind turbine

Fig. 9.5 *Left*: Dinghy for diving within *alpha ventus*. Divers checking, lying up, and obtaining measuring instruments in test district *alpha ventus*. *Right*: Construction boat for offshore wind farm mounting close to *alpha ventus*

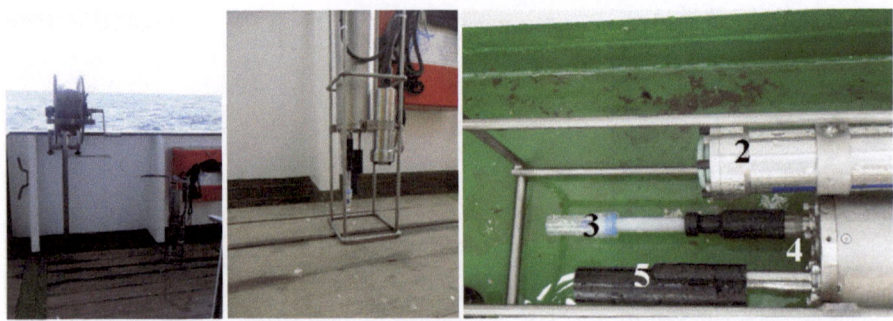

Fig. 9.6 Winch and CTD probe (*left*) and sensing elements (*middle* and *right*). Number through images: (*1*) data transfer and power supply cable and connection and rope of winch, (*2*) oxygen sensor, (*3*) temperature probe, (*4*) pressure sensor, (*5*) conductivity cell

This means that, connected with small temperature coefficient, 2.5×10^{-6} parts of the resistance value must be resolved. Therefore, Pt100 is used as resistance within a Wheatstone bridge circuit.

Conductivity Measurement
The used technique for conductivity is a galvanic method with two electrodes via a bridge circuit. The electrodes are sensitive against exterior electrical and magnetic fields. They have to be protected towards outside then the surrounding cannot take an influence on the measuring signal.

Pressure Measurements
Pressure measurements with electrical technique are done by diaphragm capsule. The bending of a membrane determined the measurement. The bending is gathered piezoresistive.

In sum, 42 CTD profiles at different positions were taken by cranking the CTD probe from sea surface to bottom and back during the WEGA cruise 141 on 12 May 2013.

9.1.3 ADCP

Figure 9.7 shows the instrument ADCP (acoustic Doppler current profiler) with it is four sensors and construction for placement at sea ground. The ADCP is fixed within a ground track, which again is connected to some weight via a cable to make sure that the rack will be placed at ground. The connected buoys are swimming at surface and mark the position of dropped ADCP.

ADCPs are used to measure how fast water is moving across an entire water column, that is, water velocity over a profile. The water currents are measured via sound considering the Doppler effect. The ADCP transmits pings of sound at a constant frequency into the water from bottom to surface. As the sound waves

Fig. 9.7 Measuring instrument ADCP with *yellow* protecting above sensors (*left*) and ADCP with *four red* transducer faces within ground rack (*middle*); *right image* shows construction of ADCP ground rack (*1*), weights (*2*), and buoy (*3*) connected by cable rope (*4*) before dropping

travel, they ricochet off particles suspended in the moving water and reflect them back to the instrument. Due to the Doppler effect, sound waves bouncing back from a particle moving away from the profiler have a slightly lowered frequency when they return. Particles moving toward the instrument send back higher frequency waves. Then the difference in frequency between the waves the profiler sends out and the waves it receives is called the Doppler shift. The shift is used to calculate how fast the particle and the water around it are moving. But ADCP does not send a single wave but several pulses—broadband technology. So, finally, it is not the difference of frequency between the emitted wave and the reflected wave that is measured but the variation of phase between several reflected pulses. The depth of velocity through the measured profiles is calculated by considering the time of return of the wave and the speed of sound. The column of water is partitioned into vertical elements (bins), and the ADCP "listens" to the reflected echoes at different time intervals, which correspond to given depths.

During the WEGA cruise 141, three ADCPs were sunk into the North Sea at three different positions around the wind farm *alpha ventus*. They were dropped in the morning of 11 May 2013 and collected in the morning of 13 May 2013. At 08:30 UTC, all three instruments started measures on 11 May 2013.

The ADCP worked with a broadband of 614.4 kHz with 50 pings per ensemble, a time per ping of 12.0″, thus an ensemble interval of 600.0 s and a bin size in the vertical of 0.5 m, so 60 vertical elements.

The postprocessing of ADCP data included a filtering of the tidal signal. That was necessary to actually have a change to detect upwelling and downwelling structures in the data set because, here, tides impose the dominant signal on each dynamical measurement. The done tide filtering followed the principle of the harmonic analysis.

The ADCP measurements are illustrated in Fig. 9.8. On the one hand, data are shown, including the tidal signal; on the other hand, the tidal signal is filtered by the harmonic analysis. The order of vertical velocities, after tidal extraction, is higher than in the case of model simulation, shown in Fig. 9.9. For most of the time, the

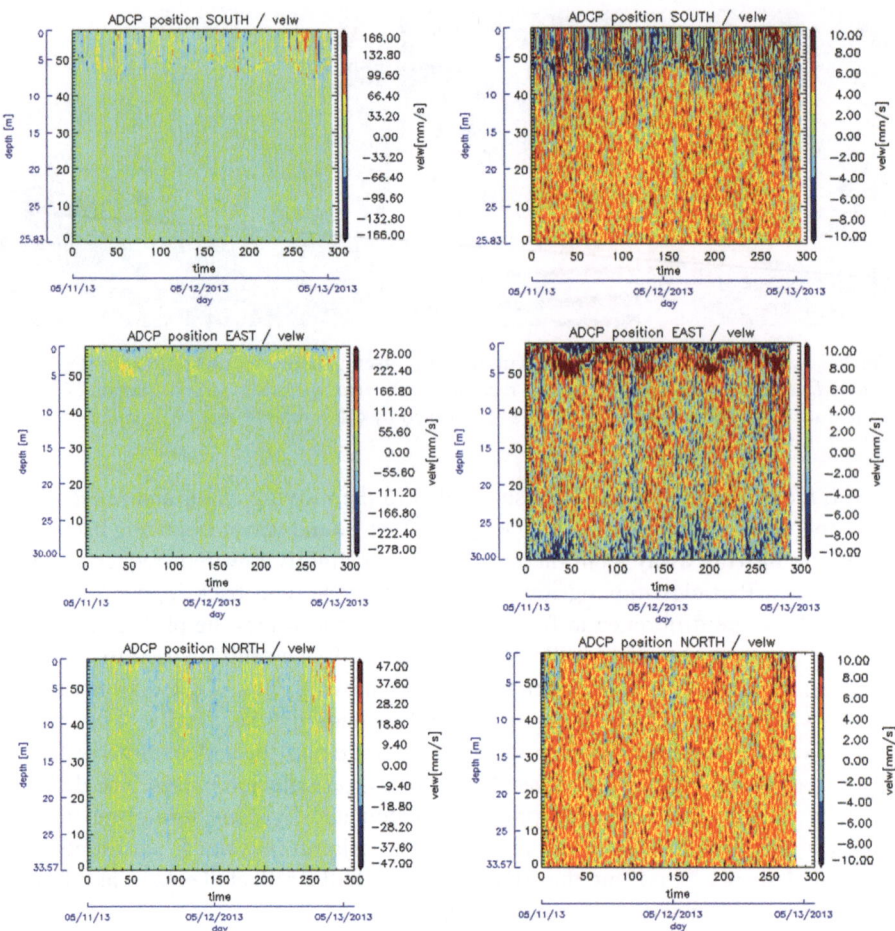

Fig. 9.8 Vertical velocity component w in mm/s for the three ADCPs around wind farm alpha ventus with tidal signal (*left*) and without tidal signal (*right*). *Black colored x*-axis and *y*-axis give time and depth at measure index. *Blue colored x*-axis and *y*-axis give the corresponding real depths and times for the measure index

measurements show a positive vertical component (greenish, reddish color gives positive values in Fig. 9.8). In connection with the analysis of the CTD temperature profiles, in Sect. 5.4, the used ADCP positions seem to be also outside of the downwelling region, which is expected within and close around the wind farm alpha ventus, but they even do not strike the upwelling region, shown in Fig. 9.9, by means of modeled velocity component w.

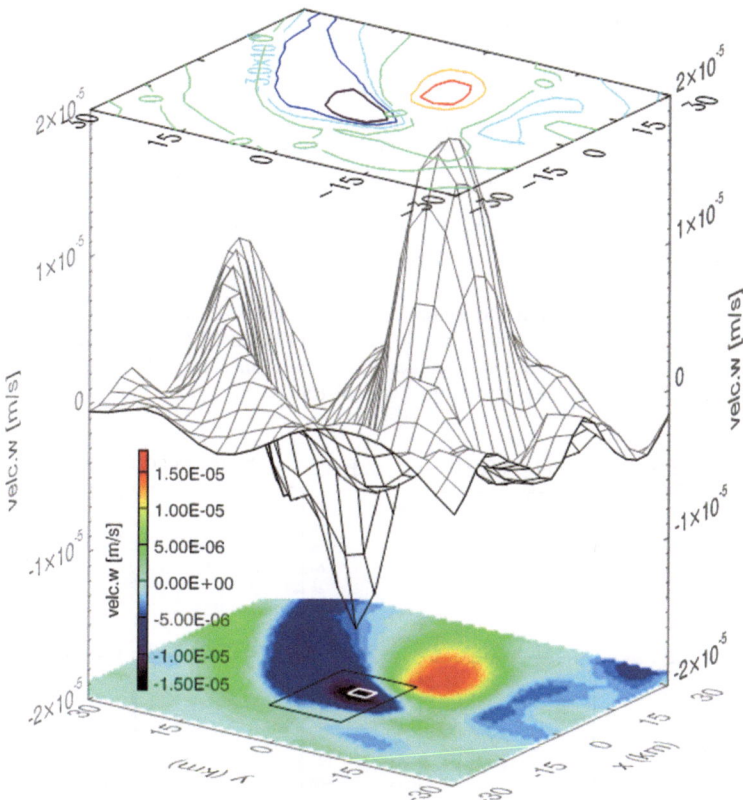

Fig. 9.9 Simulated vertical velocity component w with the ocean model HAMSOM at 2-m depth. *White square* shows the position of the OWF; *black square* illustrates the *big square* around the wind farm after modification of horizontal dimension in Sect. 5.4

9.2   Comment on Result Presentation

The analysis of model results and measurements focuses, on one hand, on the whole model area and, on the other hand, along sections through the model area and on special points. With the following graphics, positions of OWFs and investigation locations, which are used in the analysis, are clarified. Graphic Fig. 9.10 shows how small the different OWF districts are. Graphic Fig. 9.11 underlines cross sections through OWF, which were mostly used to illustrate the OWF's effect on the ocean in the vertical. Lastly, graphic Fig. 9.12 documents positions of location within the model area. Position $P0$, respectively $P3$, is placed within OWF at southeasterly grid box of the 12-turbine OWF district. Distance to S–N cross section of positive positions ($P+...$) is 6 km; for negative positions ($P-...$), it is 12 km. $P2$ is 6 km North to OWF center along S–N section; $P3$, 12 km South to it (Fig. 9.12 a and b).

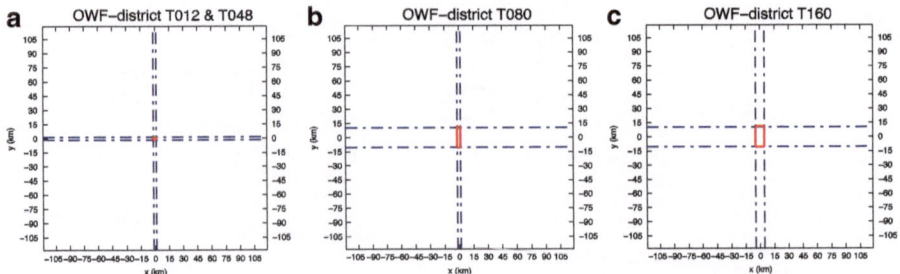

Fig. 9.10 OWF districts within the model area under TOS-01 (ocean box model), which is marked with a *red square*. Illustration (**a**) clarifies the OWF district for 12 and 48 turbines, (**b**) for 80 turbines, and (**c**) for 160 wind turbines. The *blue dashed–dotted lines* are used in result presentation; their enclosed area defines the OWF district

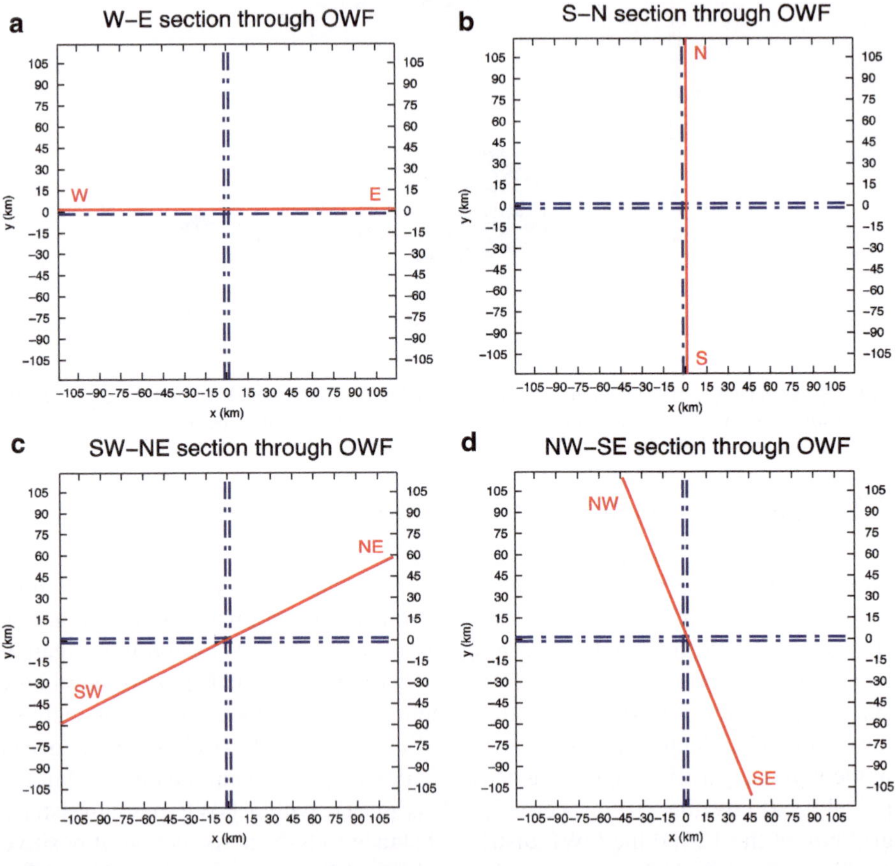

a W–E section through OWF
b S–N section through OWF
c SW–NE section through OWF
d NW–SE section through OWF

Fig. 9.11 Cross sections of analysis through OWFs. Cross section through OWF (**a**) from west to east, (**b**) from south to north, (**c**) from southwest to northeast, and (**d**) from northwest to southeast

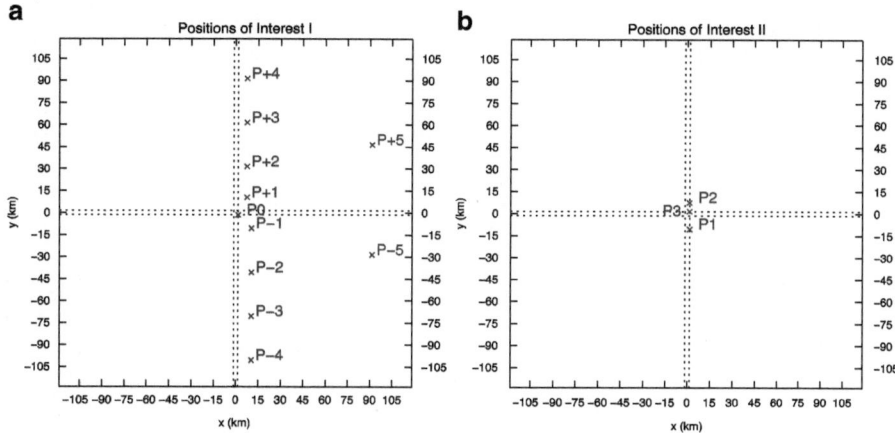

Fig. 9.12 Positions of interest for OWF analysis. (**a**) Sample of positions based on OWF induced dipole structure of surface elevation. (**b**) Sample of positions along S–N cross section through OWF based on location of extrema of ocean variables

About the International Max Planck Research School for Maritime Affairs at the University of Hamburg

The International Max Planck Research School for Maritime Affairs at the University of Hamburg was established by the Max Planck Society for the Advancement of Science, in cooperation with the Max Planck Institute for Foreign Private Law and Private International Law (Hamburg), the Max Planck Institute for Comparative Foreign Public Law and International Law (Heidelberg), the Max Planck Institute for Meteorology (Hamburg), and the University of Hamburg. The School's research is focused on the legal, economic, and geophysical aspects of the use, protection, and organization of the oceans. Its researchers work in the fields of law, economics, and natural sciences. The School provides extensive research capacities, as well as its own teaching curriculum. Currently, the School has 22 directors, who determine the general work of the School, act as supervisors for dissertations, elect applicants for the School's Ph.D. grants, and are the editors of this book series

Prof. Dr. Dr. h.c. mult. Jürgen Basedow is the director of the Max Planck Institute for Foreign Private Law and Private International Law; *President and Professor Monika Breuch-Moritz* is the president of the German Federal Maritime and Hydrographic Agency; *Prof. Dr. Dr. h.c. Peter Ehlers* is the director ret. of the German Federal Maritime and Hydrographic Agency; *Prof. Dr. Dr. h.c. Hartmut Graßl* is director emeritus of the Max Planck Institute for Meteorology; *Dr. Tatiana Ilyina* is the leader of the research group "Ocean Biogeochemistry" at the Max Planck Institute for Meteorology in Hamburg; *Prof. Dr. Florian Jeßberger* is the head of the International and Comparative Criminal Law Division at the University of Hamburg; *Prof. Dr. Lars Kaleschke* is a junior professor at the Institute of Oceanography of the University of Hamburg; *Prof. Dr. Hans-Joachim Koch* is director emeritus of the Seminar of Environmental Law at the University of Hamburg; *Prof. Dr. Robert Koch* is the director of the Institute of Insurance Law at the University of Hamburg; *Prof. Dr. Doris König* is the president of the Bucerius Law School; *Prof. Dr. Rainer Lagoni* is director emeritus of the Institute of Maritime Law and the Law of the Sea at the University of Hamburg; *Prof. Dr. Gerhard Lammel* is a senior scientist and Lecturer at the Max Planck Institute

© Springer International Publishing Switzerland 2015
E. Ludewig, *On the Effect of Offshore Wind Farms on the Atmosphere and Ocean Dynamics*, Hamburg Studies on Maritime Affairs 31,
DOI 10.1007/978-3-319-08641-5

for Chemistry, Mainz; *Prof. Dr. Ulrich Magnus* is the managing director of the Seminar of Foreign Law and Private International Law at the University of Hamburg; *Prof. Dr. Peter Mankowski* is the director of the Seminar of Foreign and Private International Law at the University of Hamburg; *Prof. Stefan Oeter* is the managing director of the Institute for International Affairs at the University of Hamburg; *Prof. Dr. Marian Paschke* is the managing director of the Institute of Maritime Law and the Law of the Sea at the University of Hamburg; *PD Dr. Thomas Pohlmann* is a senior scientist at the Centre for Marine and Climate Research and Member of the Institute of Oceanography at the University of Hamburg; *Dr. Uwe A. Schneider* is an assistant professor at the Research Unit Sustainability and Global Change of the University of Hamburg; *Prof. Dr. Detlef Stammer* is a professor in Physical Oceanography and Remote Sensing at the Institute of Oceanography of the University of Hamburg; *Prof. Dr. Jürgen Sündermann* is director emeritus of the Centre for Marine and Climate Research at the University of Hamburg; *Prof. Dr. Rüdiger Wolfrum* is director emeritus at the Max Planck Institute for Comparative Foreign Public Law and International Law and a judge at the International Tribunal for the Law of the Sea; *Prof. Dr. Wilfried Zahel* is professor emeritus at the Centre for Marine and Climate Research of the University of Hamburg.

At present, *Prof. Dr. Dr. h.c. Jürgen Basedow* and *Prof. Dr. Ulrich Magnus* serve as speakers of the Research School.